# MATERIALS SCIENCE AND
# ENGINEERING LAB MANUAL

## The PWS Series in Engineering

# MATERIALS SCIENCE AND ENGINEERING LAB MANUAL

SHERIF D. EL WAKIL

University of Massachusetts Dartmouth

**PWS Publishing Company**
**Boston**

PWS PUBLISHING COMPANY
20 Park Plaza, Boston, MA 02116-4324

PWS Publishing Company is a division of Wadsworth, Inc.

 TM

International Thomson Publishing
The trademark ITP is used under license

 *This book is printed on recycled, acid-free paper.*

```
┌─────────────────────────────────────────────────────────────┐
│          Library of Congress Cataloging-in-Publication Data   │
│ El Wakil, Sherif D.,                                          │
│      Materials science and engineering lab manual / Sherif D. El Wakil │
│      p.    cm.                                                │
│      Includes bibliographical references and index.          │
│      ISBN 0-534-93417-X                                       │
│      1.  Metallography–Laboratory manuals.    I.   Askeland, Donald R. │
│ II   Title.                                                   │
│ TN690.4.E4   1993                                 93–27513    │
│ 669′.95—dc20                                          CIP     │
└─────────────────────────────────────────────────────────────┘
```

**Photo Credits:** (Figure 1–2, p. 8; Figure 1–5, pp. 12–13; Figure 1–10, p.18; Figure 1–11, p. 19;
Figure 1–13, p. 20; Figure 1–15, pp. 22–24; Figure 1–17, p. 27—all Courtesy of Buehler Ltd.)
(Figure 2–1, p. 52—Courtesy of Tinius Olsen Testing Machine Co., Inc., Willowgrove, PA.)

Sponsoring Editor: Jonathan Plant
Production Editor: Monique Calello
Assistant Editor: Mary Thomas
Editorial Assistant: Cynthia Harris
Marketing Manager: Nathan Wilbur
Manufacturing Coordinator: Ruth Graham
Interior Designer: Julia Gecha/M. Calello
Interior Illustrator: George Nichols
Cover Designer: Julia Gecha/M. Calello
Cover Photo/Art: W. Cody/Westlight
Compositor: Pine Tree Composition, Inc.
Printer and Binder: Malloy Lithographing

Printed and bound in the United States of America.
93 94 95 96 97 98 — 10 9 8 7 6 5 4 3 2 1

To Fatima, Omar, Karim, and Sherif Jr.

# Contents

# Preface

The idea for this manual is a product of twenty years of teaching, during which it was clearly evident that the absence of a laboratory manual caused many problems for students and instructors alike. While there are plenty of traditional materials science textbooks available on the market, they are not of any significant help in the laboratory because they do not explain the operations of the equipment used in the laboratory, do not cover the fundamentals of metallographic procedures and techniques, and lack any description of appropriate laboratory experiments. Classroom handouts are also inadequate because they cannot cover topics like metallography in the depth necessary for university-level engineering students.

For these reasons, the main goal of this manual is to provide adequate coverage of the experimental techniques and procedures used in metallography as well as the working principles of the equipment typically used in the materials science laboratory.

The text is divided into three parts. The first part includes a comprehensive coverage of metallographic laboratory practice and equipment operations principles. The second part describes a large group of experiments designed to support lecture materials. These experiments are explained in detail and can be performed in any reasonably equipped materials science laboratory. Whenever appropriate, questions are provided to help the student to better understand the experiment being discussed and also as an aid in the writing of laboratory reports. The third part of the text includes supporting materials, tables, and a bibliography for further reading.

## Acknowledgments

I would like to thank my wife for her patience and understanding. I would also like to thank the corporations that provided the excellent photographs that illustrate the manual. Acknowledgment must also go to Ms. Francine Gilbert for typing the manuscript and to everyone at PWS Publishing Company for their help and encouragement. I wish to acknowledge the helpful reviews of the manuscript offered by Dr. Alan Wolfenden at Texas A & M University and Dr. Ernest G. Wolff at Oregon State University.

*Sherif D. El Wakil*

# PART I

## Metallographic Laboratory Practice

## 1–1  Safety Rules for the Materials Science Laboratory

There are some procedures and precautions that should be faithfully observed to ensure the safety and security of students in the materials science laboratory. These include what a student must do in case of an emergency situation such as a fire or an explosion, as well as preventive measures that, if taken, can minimize or eliminate the possibility of accidents. Students should follow these safety rules:

1. You should be aware of the locations of the emergency switch buttons, which can be used to cut off all electric circuits in the laboratory rooms, except the lights.

2. You should be aware of the locations of fire extinguishing equipment and fire alarms in the laboratory.

3. You should write the university emergency telephone number on a piece of paper that you can keep handy at all times.

4. Any malfunctioning of laboratory equipment should be promptly reported to your instructor, as should plugged drains or other safety hazards.

5. You should be aware of the location of the first aid kit and know how to use the emergency eye wash (and shower) in the materials science lab.

6. You should not work in the materials science laboratory unless an instructor or qualified technician is present.

7. Minor cuts or burns should not be neglected. You should report them immediately to the instructor, who will decide whether to carry out first aid procedures or to send you to the university medical center for treatment.

8. You must wear safety glasses for eye protection when you are operating a grinding wheel, a power saw, or a cut-off wheel. Some precaution must also be taken when using metal cutting tools such as chisels and hammers, and when handling acids or caustic solutions. Should a droplet of any of these liquids get into your eye, promptly and continuously flood the eye with water and seek assistance from the instructor.

9. Because a materials science laboratory contains many hazardous corrosive and/or poisonous liquids, you should never drink from the laboratory glassware.

10. The alcohol that is available in the laboratory for preparing specimens for metallographic examination is denatured methyl alcohol, a highly poisonous liquid that should never be drunk or swallowed.

11. Take care to avoid spilling any liquids on the floor or other flat surfaces in the laboratory.

# 1-2  Laboratory Report Preparation

The ability to communicate clearly both orally and in writing is of great importance to professional engineers. After graduation, you will spend a good part of your time explaining your ideas and points of view both to your superiors and to the technicians under your supervision. You are, therefore, advised to work on improving your written communication skills, and the preparation of laboratory reports provides you with an excellent opportunity to do so. Unlike a newspaper article, a technical report has a certain standard format that must always be adhered to. In addition, the style of technical and scientific writing is different from that used in, say, books of literature. Colloquial language must never be used; instead, explain your ideas in simple and grammatically correct English using clear and short sentences. Try to get to the point directly and avoid unnecessary elaborations and wordy text. Also, it is a well-established tradition in scientific writing to report in the third person (i.e., avoid using "I" and "we") and in the past tense. Let us now discuss in detail the standard technical report format.

In most cases, a technical report or laboratory report should consist of the following components:

1. *Title page*    The title page should indicate the university, the college, and the department in which the materials science laboratory course is offered. It should also show the title and the serial number of the experiment performed, your name, the name of the laboratory instructor to whom the report is submitted, and the date on which the report is submitted.

2. *Objective of the experiment*    Briefly state the purpose or the goals of the experiment. This part of the report should not exceed two fairly short paragraphs.

3. *Equipment*    List the equipment used in the experiment and briefly mention the specific characteristics of each (e.g., capacity, accuracy of reading).

4. *Experimental Procedure*    Write in your own words the actual experimental procedure followed. The procedure should reflect the facilities available at the laboratory where the experiment was conducted and may, therefore, differ slightly from the standard procedure mentioned later in this manual.

5. *Experimental results*    In this part of your report, raw data taken during the course of the experiment should be presented in an appropriate form such as tables, graphs, figures, or photographs. These must be numbered and supplemented by captions that briefly describe what the figures or photographs are all about. Whenever graphics are used, special attention should be given to the drawing scale in order to yield meaningful curves that clearly indicate the significance of the results. Miniature and over-enlarged graphs should be avoided. Finally, remember to plot the independent variables (such as time or the temperature at which a sample is soaked) along the abscissa, and the dependent variable (such as the hardness of a sample after thermal treatment) along the ordinate.

6. *Discussion of results*    This part of the report should constitute a discussion of the experimental data obtained as well as the possible sources of experimental error and in what way these may have affected the results. You can also include a correlation

and/or a comparison of your experimental results with those that can be predicted from theories or the use of theoretical analysis, and try to explain any discrepancies.

7. *Conclusions*   In this section, you present what you can conclude from the results of the experiment and the analysis applied in the previous discussion. It may also be appropriate to add your personal opinion whether the experiment served its goal of reinforcing what was covered in the materials science lectures.

8. *References*   Here you provide a numbered list of scientific books or articles to which you refer in the text. Make sure that each reference was actually cited and numbered in the text, that each number corresponds to the correct reference, and that the numbers are in sequential order. You can also provide a bibliography, which is a general list of supporting materials that were not specifically referred to in the text. There is a standard format for writing a reference; use the bibliography at the end of this manual as a guideline.

# 1–3  The Use of Standards

During an experiment, or a test that is carried out to determine the physical, chemical, or mechanical properties of a material, there are always variables whose magnitudes affect the readings taken and, consequently, the results obtained. Experimental results would not, therefore, be meaningful unless the variables affecting the test are fully specified. As a consequence, it is obvious that the results obtained from two tests can be compared only when the corresponding variables in both tests are set to be identical. This leads to the need to standardize the specifications and the methods of testing, so that the results of identical tests can be compared.

The process of specifying the procedure and the parameters affecting an experimental test is not an easy one. It should be, and actually is, based on striking a balance between accuracy and practicality. Needless to say, the targeted or desired accuracy and the availability of facilities differ from one country to another, and so do the standards for the methods of testing. It is for this reason that engineering societies, governmental institutes, and manufacturers' associations in each country work together by establishing committees of specialists to compile standards for each commonly used test that should be adhered to whenever that test is carried out.

In the area of material testing and metallurgy, we in the United States should always follow the standards developed by the American Society for Metals (ASM) and the American Society for Testing and Materials (ASTM). For each test, there is a publication in which the standard conditions for performing that test are precisely specified. In the publication detailing the standards for a tension test, for example, the shape, dimensions, and surface finish of the test specimen are accurately covered. If you use a specimen having a shape or dimensions different from the standard ones, the results you obtain would not truly indicate the properties of the material tested, and could not be compared with the data given in handbooks. As you may have expected, each of these publications has an alphanumeric code or designation to which you should refer when ordering the publication or when writing a test report. A brief summary of the standards for commonly used tests is given in Part 3 of this manual.

# 1–4 Hardness Testing Methods

Hardness can be defined as the resistance of a material to scratching, abrasion, or penetration. Obviously, any hardness index is a manifestation of the combined effect of several related properties, which may include the yield point, ultimate tensile strength, malleability, work hardening characteristics, wear-resistance properties, etc. Therefore, hardness measurements must be interpreted with caution and full consideration of their attendant limitations. In fact, past methods for determining hardness, like the file hardness test, were not fully reliable because they were dependent on the skill of the technician performing the test. It was not until 1900 that Dr. Brinell of Sweden proposed a new, reliable method whereby the hardness of a metal could be indicated by its resistance to indentation. Nowadays, hardness testing has found extensive industrial applications and is an essential tool in the quality control of metals, alloys, and metal products. The most commonly used hardness testing methods are the Brinell and the Rockwell hardness tests, which we will discuss in detail. Other, specialized hardness tests are, we believe, beyond the scope of this book.

**1–4–1  Brinell Hardness Test**    This test involves forcing a hardened steel ball into the metal specimen under a definite static load, then measuring the size (diameter) of the impression produced by that ball penetrator. In this case, the hardness index, which is called the Brinell Hardness Number, is the static load acting on the penetrator divided by the spherical area of the impression, the unit being kilogram force per square millimeter. According to ASTM specifications (Designation: E 10), the penetrator used in a standard Brinell hardness test is a spherical ball having a diameter of 10 mm. These standards also specify the load as well as the duration of its application, namely 3000 kg and at least 10 seconds for ferrous metals and 500 kg and at least 30 seconds for nonferrous metals. The diameter of the impression is usually obtained by optical magnification projected on a screen, where it can be measured accurately using a vernier calliper. Alternatively, a special measuring-type microscope is sometimes used. In either case, a number of measurements must be made across different diameters. These measurements are then averaged in order to obtain a value that truly represents the diameter of the impression. The Brinell Hardness Number corresponding to this value can be obtained from tables, thus eliminating the need for calculations.

It must always be borne in mind that the ball penetrator will inevitably undergo elastic deformation when forced into the metal specimen during the test. For any specific applied load, say 3000 kg force, the magnitude of that deformation will depend on the resistance of the metal specimen to penetration (i.e., its hardness) and, obviously, on the hardness of the ball penetrator as well. The higher the hardness of the metal specimen, the more sensible the amount of deformation in the ball would be. Ordinary high carbon steel ball penetrators can only be used when the hardness of the metal specimen does not exceed a Brinell Hardness Number of about 500. For higher values of hardness (up to Brinell Hardness Number 700), the use of a tungsten carbide ball penetrator is recommended.

In medium-hard ferrous metals, there is usually a raised ridge of metal around the impression caused by the penetrator, while in brass and bronze the impression is surrounded by a depressed surface. Consequently, the diameter of the impression appears slightly larger in brass or bronze than what it really should be, as indicated in Figure 1–1. Because of that undesirable secondary plastic deformation (and work hardening), hardness measurements should not be taken close to the edges of the specimen or close to each other. The minimum distance between the center of the impression and the edge of the test specimen should be 2.5 times the diameter of the impression. Also, as a precaution in this test, the thickness of the test specimen must not be less than 6 mm (0.25 in.), so that the anvil's support will not have any influence on the penetration, a problem that results in erratic hardness values. For thinner test specimens, lighter loads and smaller penetrators are used. Finally, the condition of the test surface specified in the standards must be adhered to. The surface of the specimen must be flat, reasonably smooth, and free from defects in order to obtain meaningful results.

As we will see later, the obvious advantages of the Brinell hardness testing method are that only one theoretical linear scale of hardness is used, regardless of the metal being tested, and that the hardness of the metal on that scale and its ultimate tensile strength are indeed correlated.

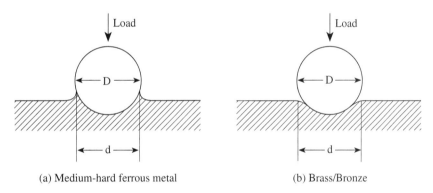

(a) Medium-hard ferrous metal            (b) Brass/Bronze

**FIGURE 1–1**    Schematic illustration of the Brinell hardness test.

**1–4–2  Rockwell Hardness Test**    Similar to the Brinell hardness test, the Rockwell hardness test involves forcing an indentor into the test specimen under static load. However, in the Rockwell testing method, the hardness index is determined by measuring the increment of depth (of the impression) as a result of applying a primary and a secondary load, instead of measuring the diameter. Consequently, there is no need for optical measurements or calculations and the Rockwell hardness number is readily shown on a dial indicator. The Rockwell hardness test is, therefore, commonly used in industry because of its simplicity and the ease with which it can be performed. Figure 1–2 illustrates a Rockwell hardness testing apparatus.

There are basically two standard indentors that are used with two hardness scales to determine the Rockwell hardness numbers for nearly all the common metals and alloys (ASTM Designation: E 18). These indentors (penetrators) are a hardened steel ball hav-

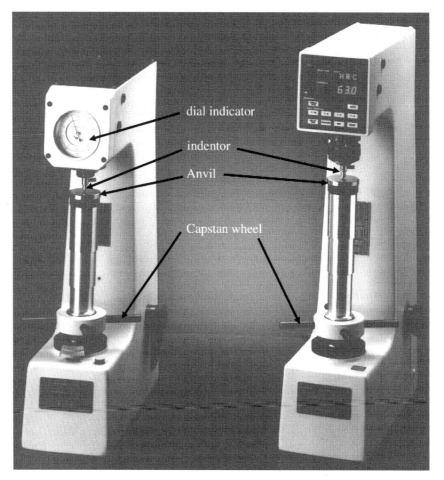

**FIGURE 1-2**   A photograph of a Rockwell hardness testing apparatus. *(Courtesy of Buehler Ltd.)*

ing a diameter of 1.59 mm (1/16 in.) and a diamond cone having an apex angle of 120° and a rounded tip 0.2 mm in radius (called the Brale), and are used with scales designated as B and C, respectively. The working range of scale B, which is used for nonferrous metals and annealed low carbon steels, is from Rb 0 to Rb 100. For the sake of measurement accuracy, when the hardness of the material being tested exceeds Rb 100, you must switch to scale C; if the hardness is less than Rb 0, another appropriate Rockwell hardness scale should be used. The useful range of the C scale, which is used for hardened and tempered steels, is from Rc 20 (equivalent to Rb 97) to slightly above Rc 70. Owing to inherent inaccuracies associated with shaping the Brale, the C scale should not be used for measuring hardness below Rc 20; instead, the hardened steel ball and scale B are usually employed.

Figure 1–3 shows the procedure for performing a Rockwell hardness test on the C scale. First the test specimen is placed on the anvil at the upper end of the elevating screw. The capstan wheel is then rotated so as to bring the surface of the test specimen

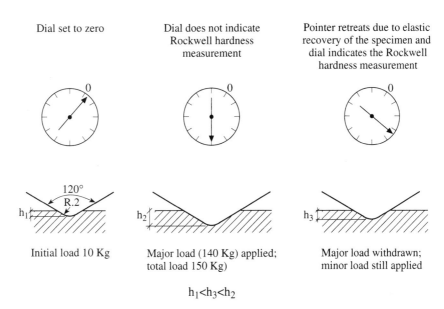

| Dial set to zero | Dial does not indicate Rockwell hardness measurement | Pointer retreats due to elastic recovery of the specimen and dial indicates the Rockwell hardness measurement |
| Initial load 10 Kg | Major load (140 Kg) applied; total load 150 Kg) | Major load withdrawn; minor load still applied |

$$h_1 < h_3 < h_2$$

**FIGURE 1-3**    The procedure for performing a Rockwell hardness test (on the C scale).

in contact with the penetrator. By further rotation of the wheel, the test specimen is forced against the indentor and a minor load of 10 kg is slowly applied in order to seat the specimen firmly. At this moment, the dial indicator of the apparatus (whether mechanical or optical) is set to zero. Next, an additional load of 140 kg (90 kg in a test on the B scale) is applied by means of a release handle mounted on the side of the apparatus. The total major load will now be 150 kg and the duration of its application should be at least 10 seconds. Obviously, the application of that load would force the penetrator into the specimen to an additional depth. Still, that depth must not be considered as an indication of hardness because it includes an elastic as well as a plastic deformation. Therefore, the additional load is released without removing the minor load, and the hardness index is then shown on the dial indicator. That reading reflects the permanent or plastic increment of penetration depth resulting from the increment of load between the minor and major loads. It does not indicate the total depth of penetration of the indentor. Again, as in the case of the Brinell hardness test, care must be taken to ensure that the surface conditions of the test specimen—its flatness and its thickness—are within the limits specified by the standards.

### Review Questions on Hardness Testing

1. How do you define hardness?

2. What is the theory on which Brinell hardness testing is based?

3. What are the advantages of the Brinell hardness testing method?

4. What is the theory on which Rockwell hardness testing is based?

5. What is the purpose of the minor load in Rockwell hardness testing?

6. What are the advantages of the Rockwell hardness testing method?

7. What are the disadvantages of the Rockwell method?

8. How thick must the specimen be in order to get an accurate Rockwell hardness reading?

9. How close can a Rockwell hardness indentation be to another without getting an error in the reading?

10. What is the main difference between a regular hardness tester and a micro-hardness tester?

# I–5  The Metallurgical Microscope

The optical microscope is the tool that is commonly used for examining and photograph-ically recording the microstructures of metals and alloys. Since metals and alloys are always opaque and do not allow light to pass through, no matter how thin they are, the metallurgical microscope differs from the biological type in the manner by which the specimen under investigation is illuminated. As can be seen in Figure 1–4, which indi-cates a schematic illustration of the metallurgical microscope, a special external source of light (called the illuminator) is employed. It is also evident that the metallurgical microscope is composed of two distinct and separate optical lenses, namely the objec-tive and the eyepiece. In fact, we will see later that each one of them can, and usually is, replaced by an optical system of lenses in order to improve the clarity with which the microscopic image is observed.

Let us now discuss the principles of operation of the metallurgical microscope. A horizontal bundle of rays of light is emitted from the illuminator and diverted down-ward, by means of a plane glass reflector to pass through the objective and fall onto the specially prepared surface of the metal specimen. The light that is reflected from the surface of the specimen will pass again through the objective to form an enlarged pri-mary image of the illuminated area. The eyepiece is then used to further magnify the

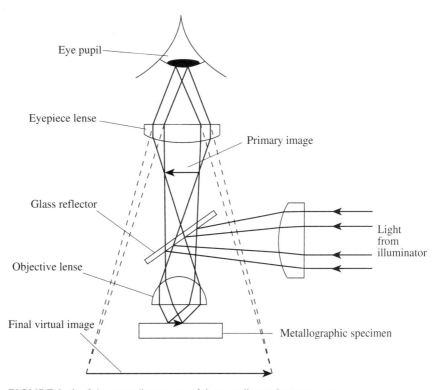

**FIGURE I–4**   Schematic illustration of the metallurgical microscope.

primary image for visual examination or to project it onto a photographic film so that it can be permanently recorded. The power of magnification of the microscope is the product of the powers of magnification of the objective and the eyepiece. In a modern metallurgical microscope, like those shown in Figure 1–5, the objective can be changed easily in order to obtain a higher or lower magnifying power. Magnifications of up to 1000× can be obtained and are quite common, but the clarity of the obtained image decreases at higher magnifications.

In order to be corrected for two defects occurring in simple lenses, the objective must be composed of an optical system of lenses. These two defects are the chromatic and the spherical aberrations and are responsible for destroying sharp definition in the image. Chromatic aberration occurs because white light is actually a mixture of color lights that have different wave lengths and are, therefore, refracted by the optical glass of the lens to different degrees. As can be seen in Figure 1–6, the final outcome will be three images of different colors, the violet or blue image being closer to the lens and the red image farthest. The remedy for this problem involves the use of either an achromatic

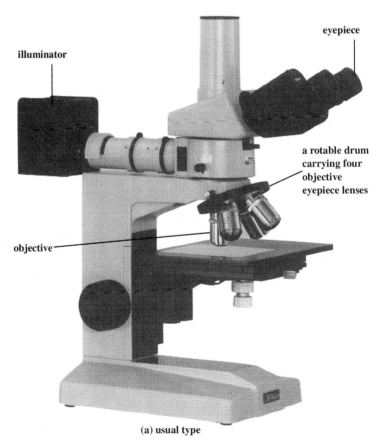

**(a) usual type**

**FIGURE 1–5**    Photographs of modern metallurgical microscopes. ((a) *Courtesy of Nikon;* (b) *Courtesy of Buehler Ltd.*)

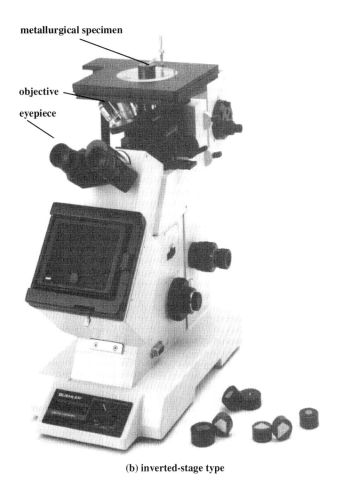

metallurgical specimen

objective

eyepiece

**(b) inverted-stage type**

**FIGURE 1–5**    (continued)

or apochromatic objective. As shown in Figure 1–7, the achromatic objective is corrected to bring green and red images to a focus in the same plane, whereas the apochromatic objective is highly corrected and brings the violet, green, and red images to a single sharp focus. The spherical aberration is caused by the fact that the light beams passing through the outermost margins of a lens are refracted to a greater degree than are

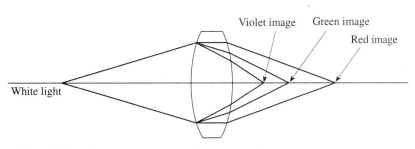

Violet image    Green image

Red image

White light

**FIGURE 1–6**    Schematic illustration of chromatic aberration.

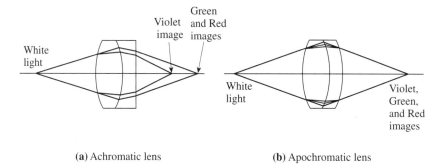

**(a)** Achromatic lens           **(b)** Apochromatic lens

**FIGURE 1–7** Using achromatic and apochromatic objectives to minimize and eliminate chromatic aberration.

identical light beams passing through the lens near its principal axis. As shown in Figure 1–8, this would result in two images, one for the marginal rays and one for the central rays. Again, the apochromatic objective provides a good remedy for this problem.

It is evident from the discussion above that the use of monochromatic light (i.e., single-color light with a definite wave length) would result in an image having the best definition possible. Therefore, a light filter that allows only a single monochromatic light is commonly placed before the objective, particularly when photographs of the microstructure are to be taken. Filters basically involve colored glass, with blue, green, and yellow being the commonly used colors. Some filters consist of a dyed gelatine film sandwiched between two thin glass slides, while others take the form of a cell filled with a colored solution.

For the sake of completion, let us now briefly discuss two terms that are frequently used in metallurgical microscopy, namely the resolving power and the numerical aperture. The resolving power of an optical system is the capacity to show lines that are very close, clearly well-defined, and separated. An important variable on which the resolving power of a lens (e.g., objective) depends is the angle of aperture, i.e., the angle confined between the extreme rays coming from the focal point to the outer edges of the lens. The higher the angle of aperture of a lens the higher its resolving power, because more light will be collected and allowed to pass through. In fact, it can be mathematically proven that the resolving power is proportional to the sine of the half of the aperture angle and

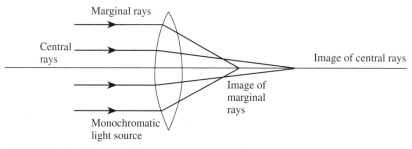

**FIGURE 1–8** Schematic illustration of spherical aberration.

to the index of refraction of the medium between the lens and the specimen. The product of these two variables is referred to as the numerical aperture and is an indication of the resolving power of a lens (or an objective optical system).

## Review Questions on The Metallurgical Microscope

1. How and why is the metallurgical microscope different from the biological type?

2. Of what two distinct optical systems is a metallurgical microscope composed?

3. Draw a neat sketch to illustrate the principles of operation of a metallurgical microscope.

4. Knowing the magnifying power of each of the two optical systems of a microscope, how can you calculate the magnifying power of that microscope?

5. What are the two defects occurring in simple lenses? In what way can they affect the proper functioning of a metallurgical microscope?

6. In a modern metallurgical microscope, how can the common defects in lenses be eliminated?

7. Why is it advantageous (especially when taking a photograph) to use a light filter?

8. What are light filters composed of?

9. What is the resolving power of an objective and what are the variables affecting it?

10. What is the angle of aperture and what effect will enlarging that angle have on the obtained image?

11. Define the numerical aperture of an objective optical system and explain why a drop of oil is sometimes used to fill the gap between the objective and the specimen.

## 1–6  Preparation of Specimens for Metallographic Examination

Metallography is basically the study of the structures and constitution of metals and alloys, using metallurgical microscopes and magnifications from 100× to 1500×, so that the physical and mechanical properties of an alloy can be related to its observed microstructure. Such microscopic studies can provide the trained engineer with an abundance of constitutional information about the specimen under investigation, including the size and shape of the grains (crystallites), the presence of microdefects (such as segregation, hair cracks, and nonmetallic inclusions), and the nature and distribution of secondary phases. The metallographic examination can, therefore, be used in quality control and to predict and/or explain the mechanical properties of an alloy to aid in determining the cause of failure of a metallic component.

As was previously mentioned, the principles of operation of the metallurgical microscope involve reflecting light from the surface of the specimen. Therefore, the preparation of that surface (usually called the microsection) so that it is free from scratches, work-hardened layers, and surface flaws is of extreme importance. As a beginner, you are advised not to seek a short cut or a means of saving time because this would yield a sample that does not present the true microstructure. After familiarizing yourself with all the stages of specimen preparation, you should try to gain insight into what affects each of these stages and then develop your own preparation procedures.

The following is a detailed discussion of the process of preparing a quality microsection for metallographic examination (ASTM Designation: E 2).

**1–6–1 Sectioning**    Because of the difficulty of and the longer times consumed in polishing large samples, the area for metallographic examination should be relatively small and usually does not exceed 12 mm × 12 mm (0.5 in. × 0.5 in.). Therefore, a truly representative sample should be selected carefully so as to provide the maximum information over the smallest area. If all the desired microstructure characteristics cannot be observed in one reasonably small specimen, then more specimens must be prepared in order to reveal all microstructures of interest. These samples for sectioning are cut from larger pieces by means of a hacksaw, a band saw, or, in the case of hardened metals, an abrasive cut-off wheel. Throughout the cutting operation, care must be taken to avoid alteration of the microstructure by excessive heating or work hardening as a result of the cutting action. The workpiece must be flooded with a coolant and the rate of cutting should be kept to a minimum. Next, you should examine the specimen visually and remove any burrs by filing or any other appropriate means. Finally, the specimen should be carefully washed before proceeding to the next stage.

For convenience in handling, small samples are usually mounted in a matrix of phenolformaldehyde resin, bakelite, plexiglass, or any other appropriate thermoplastic or thermosetting polymer. Mounting also has the advantage of protecting the extreme edges and enabling micrographs of these areas to be taken. At the same time, the polymeric matrix does not affect etching because the polymers employed are inert to alcohol and the commonly used etchants. Following is a discussion of the two most common mounting methods employed in metallurgical laboratories, namely, compression mounting and cold mounting.

a. *Compression mounting*    This method involves using pressure combined with heat to encapsulate the sample in the polymeric material by means of a special small press. Figure 1–9 shows the stages of the mount molding cycle. First, the surface of the specimen to be examined is placed facedown on the mounting die. The die is then filled with resin powder around the specimen and the assembly is subjected to heat and pressure simultaneously. The resin supplier's data should be consulted to select the appropriate magnitude of the applied pressure as well as its duration. Finally, the die is cooled (either manually by placing a heat-sink, sometimes water-cooled, around it, or automatically in modern mounting units like that shown in Figure 1–10), and the mount is ejected from the mold after releasing the pressure. It is of supreme importance that you now use a suitable sharp tool to scratch the plastic immediately with an identification mark that you have established.

b. *Cold mounting*    When heat and/or pressure can cause damage to the metallurgical specimen or change its microstructure, the cold mounting method is recommended. In this method, the polymerization reaction, required to produce a rigid matrix, takes place at room temperature and requires no pressure at all. The procedure is simple and involves thoroughly mixing the resin with its catalyst and filling a ceramic mold in which the specimen (with its surface facedown on a glass sheet) is placed. It is always a good idea to place an identification tag in the mold to avoid the need to scratch the mount, which may be difficult for this kind of thermosetting resin. In this method, epoxy resins are most commonly used and offer the advantages of adhering tenaciously to the specimen and exhibiting very low shrinkage. The only disadvantage is that some resins may require several hours to cure fully.

**1–6–2 Coarse Grinding**    In this stage, the mounted sample is subjected to grinding on a belt grinder, until all the polymeric resin is removed from the entire surface of the metallurgical sample. You must make sure that the entire surface of the sample is being ground and that the surface remains almost normal to the axis of the mount. This can be achieved by frequent changing of the direction of the sample on the belt grinder. Next,

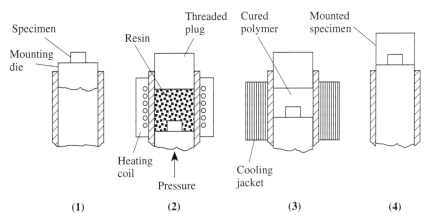

**FIGURE 1–9**    The stages of the mount molding cycle.

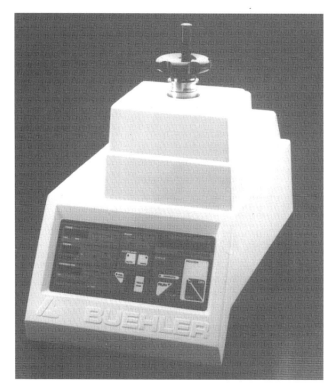

**FIGURE 1–10**   A modern mounting unit. (*Courtesy of Buehler Ltd.*)

chamfer the edge at an angle of about 45° all around to avoid ripping polishing cloths (later during the polishing stage) with sharp edges. Chamfering also has the advantage of eliminating the possibility of breaking off chips that may scratch the surface of the specimen.

**1–6–3 Fine Grinding**   As a result of the sectioning and coarse grinding operations, the surface layer of the metal specimen undergoes plastic deformation, which in turn leads to distorted crystals and altered microstructure. Fine grinding is therefore aimed at gradually reducing the thickness of that distorted layer. Fine grinding results in a newly formed deformation zone that is thinner and less severe than the original one. Since the depth of the deformation zone is directly related to the coarseness of the abrasive particles on the emery paper, the sample must be ground using progressively finer abrasive particles. Consequently, this approach gradually decreases the thickness of the distorted cold-work layer as the particle size of the abrasive decreases. For best results, a fine grinding sequence involving a series of emery papers with grit sizes of 240, 320, 400, and 600 is recommended (remember that the larger the number of the grade, the finer the particles). Figure 1–11 illustrates a typical bench unit for hand grinding metallographic specimens.

**FIGURE 1–11**    A bench unit for hand grinding metallographic specimens. (*Courtesy of Buehler Ltd.*)

The best grinding is done wet and by hand on a flat surface. A stream of water is necessary to cool down the metal sample and to wash away loosened grit, thus preventing both scratching of the surface being ground as well as clogging of the emery paper. The specimen's surface is pressed against the emery paper and the specimen is then moved along a straight line (either forward or backward) using a minimum of pressure. Also, remember to hold the sample so that the scratch marks caused by the strip of emery paper are at 90° to those formed by the preceding strip of emery paper, as shown in Figure 1–12. This makes it easier to determine when the scratches from the preceding grade of emery paper are gone. Again, it is very important that you wash the specimen and your hands after each grinding step to ensure the absolute cleanliness of the sample before going to the next finer grade of emery paper. After using the 600 grit paper and washing the specimen thoroughly, the specimen should be examined at low magnifica-

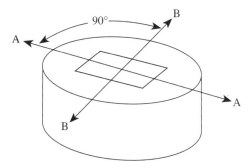

AA: direction of grinding in one step
BB: direction of grinding in the next step

**FIGURE 1–12**    Rotation of the specimen during fine grinding.

tion (10×) to make sure that the quality of the surface is appropriate before proceeding to the rough polishing operation.

**1–6–4 Rough Polishing**    This is the most important operation in the procedure for preparing metallographic specimens. It is aimed at removing from the surface to be examined the fine scratches that would result from the preceding grinding operation, thus producing a highly polished surface. Rough polishing is usually done in three stages on a set of rotating wheels similar to that shown in Figure 1–13 and covered with napless cloth such as nylon that is charged with an appropriate abrasive. Oil-soluble diamond abrasive pastes (6- and 9-micron) are well suited for this operation and yield the best results. In this case, oil (a few droplets) is added as an extender and as a lubricant. Also, suspensions of aluminum oxide having particle sizes of 15, 6 (or 3), and 1 micron are used. (Remember that each suspension must be used with the same wheel designated for it all the time, i.e., a suspension with 6-micron particles must not be charged onto a wheel designated for 1-micron particles.)

You can prepare a suspension by adding two spoonfuls of aluminum oxide (having the required particle size) to about 600 cc of distilled water and then shaking the bottle very well. Again you are reminded to thoroughly wash both the specimen and your hands between steps. Also, in order to ensure uniform surface removal throughout the entire surface, the specimen should be moved in a clockwise direction (as shown in Figure 1–14) since polishing wheels normally rotate counterclockwise.

**FIGURE 1–13**    A bench-type metallographic polishing unit. (*Courtesy of Buehler Ltd.*)

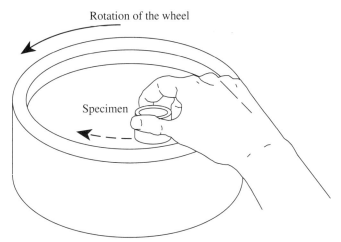

**FIGURE 1–14**    Manipulation of the specimen on a polishing wheel.

**1–6–5 Final Polishing**    After this operation, the metallographic sample must have a scratch-free surface with a mirror-like finish. Final polishing is usually conducted on a revolving wheel covered with a napped cloth (like synthetic rayon "Microcloth"® or velvet) onto which a suspension of aluminum oxide (alumina) having a submicron particle size is charged. In this operation, the specimen is manipulated with respect to the revolving wheel in the same manner as in rough polishing. Heavier hand pressure should be used in this operation, though pressure may be reduced near the end. Also, the operating time should be kept to an absolute minimum. Too long a holding period on this wheel would result in pitting of the surface, which would detrimentally affect its flatness.

It is worth mentioning here that there are three basic crystallographic forms for alumina, designated as alpha, beta, and gamma. The alpha and the gamma forms are the ones most widely used as polishing abrasives. The alpha type of alumina has a hexagonal crystal lattice, 1 $\mu$m or larger particle size, and a very high hardness. As you can see, this type of alumina is recommended for rough polishing. On the other hand, the gamma alumina is highly recommended for final polishing. It has a cubic crystal lattice, very small particle size of 0.1 or 0.05 $\mu$m, and lower hardness than the alpha type.

After the final polishing operation is completed, you should wash your sample under running water, rub it lightly with a cotton swab, flush its polished surface with alcohol, and finally dry it under a blast of hot air. Now your sample is ready for metallographic examination. The metallurgical microscope is used at a magnification of usually 100× in order to reveal structural features like the presence of porosity, hair (micro) cracks, and/or nonmetallic inclusions. The microstructure cannot be observed yet while the specimen is still in its as-polished condition, except for certain nonferrous alloys and only when polarized light is used. You must always remember that a polished sample should be stored in a dessicator and that you should not touch or scratch the polished surface. You must, therefore, not carry a polished metallographic specimen in your pocket, nor allow any hands or clothes to come in contact with the polished surface.

## 1-7 Etching of Specimens for Microscopic Examination

In order to make visible the many structural characteristics of the metal that are not revealed by the microscopic examination of the as-polished specimen (such as the grain boundaries, plastic flow of crystallites, twinning, etc.), the polished surface must be briefly etched. The etching operation involves subjecting the surface to the chemical action of an appropriate reagent under carefully controlled conditions. In multiphase alloys, different phases do not have the same solubility in the chemical reagent and would, therefore, etch at different rates. This leads to preferential attack and/or staining of one or more of those phases, which in turn establishes the structural contrast that enables the different phases of a microstructure to be identifiable under the metallurgical microscope. On the other hand, in pure metals and single-phase alloys, the etching rate is different for the various grains and is directly related to their orientation with respect to the plane of the polished surface (keep in mind that the rate of dissolutioning of any grain differs along different crystallographic planes). You can see now that the etching mechanism for pure metals and single-phase alloys is completely different from that for multiphase alloys. You should also anticipate that the rate of etching in the first case (i.e., pure metals and single-phase alloys) is much lower than that in the second case (multiphase alloys) due to the absence of the anodic reaction (this will be covered in the

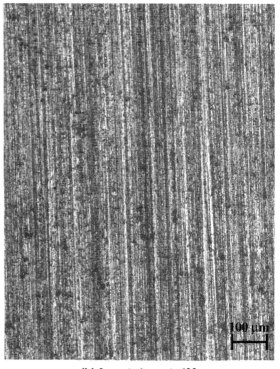

**(a)** fine grinding, grit 320              **(b)** fine grinding, grit 400

**FIGURE 1-15**   Photomicrographs of an aluminum alloy specimen after various stages of preparation for microscopic examination. (*Courtesy of Buehler Ltd.*)

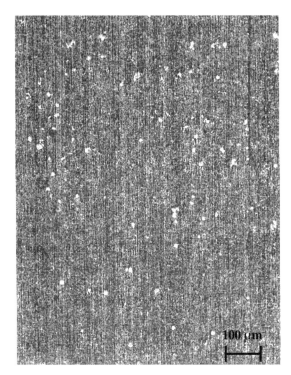

(c) fine grinding, grit 600

(d) fine grinding, grit 800

**FIGURE 1–15**    Continued

lectures on corrosion). Consequently, pure metals and single-phase alloys require longer etching time and more active reagents for their features to be adequately revealed.

Etching is conducted by either swabbing the polished surface lightly with a cotton gauze saturated with the reagent, or by immersing it into a small vessel partly filled with the reagent. In the latter case, the specimen is suspended by means of tongs in the reagent and is agitated moderately to remove the clinging air bubbles that may impede the chemical reaction. When the bright metallic shine of the polished surface disappears, the specimen is removed from the etchant and quickly rinsed with a stream of running water. The surface of the specimen is then flushed with methyl alcohol to remove water droplets and subsequently dried under a blast of warm air. If the microstructure is still not satisfactorily revealed under the metallurgical microscope, the specimen may be etched for a longer time. Nevertheless, remember not to go beyond the time recommended as this would result in an "overetched" structure that is poorly defined. Figure 1–15 shows photomicrographs of metallographic specimens after various stages of preparation for microscopic examination. The need for and importance of the etching operation can be easily realized from that figure. Naturally, certain etching reagents yield better results with certain metals than other etchants would do. A beginner is therefore advised to consult reliable references to select appropriate etchants as well as to determine the proper duration of the etching operation. For convenience, the commonly used metallographic etching reagents for ferrous and nonferrous metals are given in Table 1–1.

**(e)** polishing on Textmet® with 1-micron diamond

**(f)** polishing on Mastermet® with Microcloth®

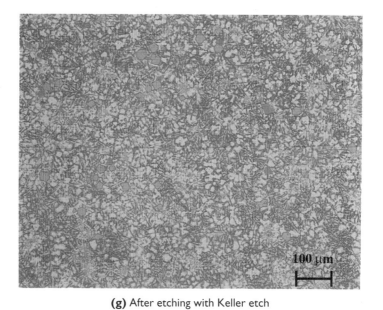

**(g)** After etching with Keller etch

**FIGURE 1–15**    Continued

**TABLE 1-1**    Commonly used etchants.

| Sample Material | Etchant | Composition | | Remarks |
|---|---|---|---|---|
| Carbon steel | Nital (usually 2%) (nitric acid) | $HNO_3$<br>Ethyl alcohol | 1–5 ml<br>100 ml | Immersion. Time ranges from a few seconds to a minute. |
| Carbon steel | Picral (picric acid) | Picric acid<br>Ethyl alcohol | 4 g<br>100 ml | Immersion. Time ranges from a few seconds to over a minute. |
| Copper, brass, bronze | Chromic acid | Saturated aqueous solution | | Immersion or swabbing 5–30 sec. |
| Copper, brass, bronze | Ammonium persulphate | $(NH_4)_2S_2O_8$<br>$H_2O$ | 10 g<br>90 ml | Immersion. Cold or boiling. |
| Aluminum | Hydrofluoric acid | HF (concen.)<br>$H_2O$ | 0.5 ml<br>99.5 ml | Swab for 15 sec. |
| Aluminum alloys | Keller's etch | HF (concen.)<br>HCL (concen.)<br>$HNO_3$ (concen.)<br>$H_2O$ | 1.0 ml<br>1.5 ml<br>2.5 ml<br>95.0 ml | Immerse for 10–20 sec. |

## 1–8 Electrolytic Polishing and Etching

As you can see from our previous discussion, the preparation of metallographic specimens by the conventional mechanical methods is a difficult and time-consuming process. Also, the complete elimination of the surface work-hardened layer is virtually impossible (it can be reduced to a minimum, though). It was mainly for these two reasons that the electrolytic polishing method was developed. As can be seen in Figure 1–16, the specimen to be polished is made the anode in an electrolytic cell, a suitable material like platinum or stainless steel serves as the cathode, and a direct current from an external source is applied through the electrolyte. As a consequence, an anodic dissolution reaction takes place, causing the peaks and ridges of the surface to dissolve faster because of the higher current density at those localities. The outcome is a smoothing and brightening of the surface, which ultimately acquires a mirror-like finish. Naturally, the process variables, i.e., the kind of electrolyte, the current intensity, and the polishing time, must be carefully selected depending on the metal of the specimen. You are advised to consult the manufacturer's manual for the apparatus being used for more details.

The process of electrolytic polishing is fairly simple and requires no great skill. After completing the fine grinding operation, the specimen is thoroughly washed with water and alcohol, loaded onto the apparatus (similar to that shown in Figure 1–17), and the current is switched on.

It is important to note that mounted specimens cannot be subjected to electrolytic polishing unless the material is electrically conductive. Also, remember that the electrolyte has to be changed (regularly) when it becomes saturated with metal ions after polishing a relatively large number of specimens.

Electrolytic etching is basically similar to electrolytic polishing, the difference being that the voltage and the current densities are considerably lower. Convenient and versatile units, which can be used for both electrolytic polishing and electrolytic etching, are commercially available. It is, however, very important to remember to wash the specimen thoroughly with water, rinse it with alcohol, and dry it with warm air after the electrolytic etching is completed.

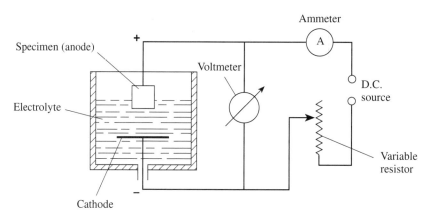

**FIGURE 1–16**   Schematic illustration of an electrolytic etching apparatus.

**FIGURE 1–17**    A modern electrolytic etching apparatus. (*Courtesy of Buehler Ltd.*)

## Review Questions on Sample Preparation and Etching

1.  Why is grinding (or polishing) conducted in more than one step?

2.  How does the quality of the metallographic finish of a specimen vary with the hardness of the abrasive used?

3.  Why is nappless cloth used in rough polishing, whereas napped cloth is used in finish polishing?

4.  Why should the edges of the piece be beveled during the coarse grinding operation?

5.  What are polishing pits and how can they be avoided?

6.  Why should the specimen be thoroughly washed after each stage during either grinding or polishing?

7.  After completion of the polishing operation, why is the specimen rinsed with alcohol after thoroughly washing it with water?

8.  Why is oil added to diamond paste during rough polishing?

9.  Why is fine grinding performed wet?

10. Why are metallographic samples sometimes mounted in a plastic matrix?

11. What are the advantages of electrolytic polishing?

12. Explain the mechanism of electrolytic polishing.

13. What is the main difference between electrolytic polishing and electrolytic etching?

# 1.9  Statistical Analysis of Data

**1–9–1 General**    In any experimental test, there is always the potential for error. Some sources of error can be readily traced and identified, while others may not be directly evident to the investigator. It is, therefore, important to be able to interpret experimental data properly and to exclude the effects of error on the results. As a matter of common sense, it is always advantageous to take more than one reading and to use the mean value as the indicative one. This is a rather simple approach, however, and it does not always work. Testing of mechanical properties often exhibits noticeable scatter, especially the testing of brittle material such as ceramics where factors like size, manufacturing method, and protection during transportation have a considerable effect. In such cases, statistical analysis of data is necessary to quantitatively study random error. Obviously, our goal is to estimate a reliable value for the mechanical property being investigated that can be used with good confidence in design. Let us now define some statistical terms and establish the approach used for the analysis of data.

**1–9–2 The Gaussian Curve**    An efficient way of summarizing raw data for easier interpretation involves grouping such data to form a *frequency distribution*. Let us illustrate this method using repeated measurements of the heights of your classmates. In a tabular form, arrange the measurements into a number of equal *class intervals (5'6"–5'8", 5'8"–6', 6'–6'2"*, etc.), and determine the frequency of the measurements in each class interval (for example, 4 students in the first, 7 in the second, etc.). Next, if the boundaries of the class interval are plotted on the x axis of a graph paper and the corresponding frequency along the y axis (to form a bar chart), a graphical representation of the frequency distribution, known as a *histogram*, is obtained, as shown in Figure 1–18. As the class interval is made narrower, we end up with a smooth bell-shaped curve like that shown in Figure 1–19, which is known as the *Gaussian frequency distribution curve*. Again, it is a common practice to express the frequency of measurements in each class interval as a percentage of the total number of measurements, so as to make the area under that curve equal to unity.

**1–9–3 Numerical Measures**    A few numerical measures are used to characterize the frequency distribution curve. The first is the simple arithmetic mean, which indicates an average value where there is a tendency for the measurements or readings to cluster. The arithmetic mean $\mu$ is given by the following equation:

$$\mu = \frac{\displaystyle\sum_{i=1}^{N} x_i}{N} \qquad\qquad 1\text{–}1$$

where $N$ is the number of measurements and $x$ is the magnitude of a measurement.

A second numerical measure is an indication of the spread or scatter of the measurements. The need for such a measure, which is called the *variance*, can be realized from

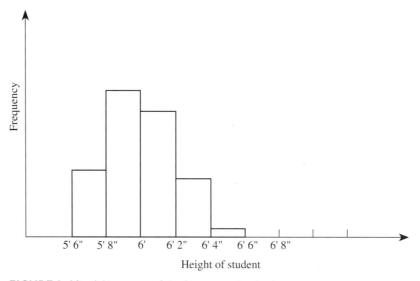

**FIGURE 1–18**   A histogram of the frequency distribution.

Figure 1–20, which indicates two curves having identical means but dissimilar frequency distributions. The variance $\sigma^2$ of a group of measurements is given by:

$$\sigma^2 = \frac{\displaystyle\sum_{i=1}^{N} (x_i - \mu)^2}{N-1} \qquad\qquad 1–2$$

Another measure of variability is the *standard deviation,* which is defined as the positive square root of the variance, and can be given in a simpler form by the following equation:

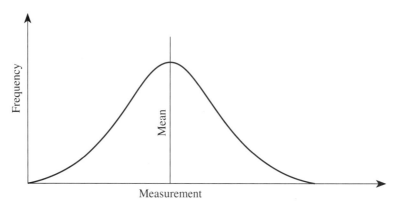

**FIGURE 1–19**   The Gaussian curve.

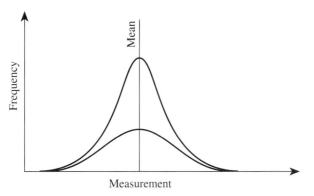

**FIGURE 1–20**   Two dissimilar frequency distributions having identical means.

$$\sigma = \left[ \frac{\displaystyle\sum_{i=1}^{N} x_i^2 - \dfrac{\left(\displaystyle\sum_{i=1}^{N} x_i\right)^2}{N}}{N-1} \right]^{1/2}$$                            1–3

**1–9–4 Probability**   Probability can be defined as the likelihood of occurrence of a reading or measurement. The Gaussian curve can be employed to determine the probability of occurrence of measurements between certain two values. Ideally, the Gaussian curve should be based on an infinite population. It is expressed mathematically in the following form (usually referred to as the *probability density function* or PDF):

$$F(x) = \frac{1}{\sigma\sqrt{2\pi}}\, e^{-(x-\mu)^2/2\sigma^2}$$                            1–4

In statistical analysis, it is always the practice to use a curve having $\mu = 0$, and a standard deviation $\sigma = 1$. That curve is called a *standard normal distribution*. In that case, the variable $x$ is called the *standardized normal random variable* and is denoted by $z$. The PDF of that newly defined variable will take the following form:

$$f(z) = \frac{1}{2\pi}\, e^{\left[-1/2(z)^2\right]}$$                            1–5

If the arithmetic mean of the PDF is not zero, as is the case in most of our applications, the PDF must be standardized by using the following algebraic substitution:

$$z = \frac{x-\mu}{\sigma}$$                            1–6

Now, considering a standard (or standardized) normal distribution like that shown in Figure 1.21, the probability that a single measurement will be greater than some value $z_2$ is given by the area under the curve to the right of $z_2$, and not by the ordinate corresponding to $z_2$ . On the other hand, the probability that a single random reading will be less than $z_2$ is given by the area to the left of $z_2$. Again, the probability that a single

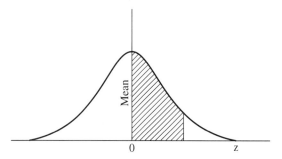

**FIGURE 1–21**   The standard normal distribution curve.

measurement will fall between $z_1$ and $z_2$ is given by the area under the frequency distribution curve bounded by normal lines at $z_1$ and $z_2$, respectively. Apparently, the total area under the curve is unity, or a probability of 100 percent. For the sake of simplicity, and in order to eliminate the need to carry out complicated calculations, tables are readily available to give the areas under the curve between $z = 0$ and different values of $z$ (see Part 3).

**1–9–5 Confidence**   Our discussion thus far has assumed analysis of a large population. Unfortunately, we usually deal with small groups, or samples, each having fewer than 30 readings. In order to accommodate small sample sizes, the following variable $t$ is introduced.

$$t = \frac{\overline{x} - \mu}{S / \sqrt{N}} \qquad\qquad 1\text{–}7$$

where $\overline{x}$ is the mean of the sample and $S$ is its standard deviation.

The $t$ distribution is symmetrical about zero and involves degrees of freedom, i.e., each different curve is dependent on the number of readings in a sample. Here, the degrees of freedom and the number of measurements in a sample are related by the following equation:

$$\gamma = N - 1 \qquad\qquad 1\text{–}8$$

where $\gamma$ is the degrees of freedom and $N$ is the number of measurements in a sample.

The probability that a given value will lie outside a specified value (or values) of $t_\alpha$ is designated by $\alpha$. Accordingly, the probability that a given value will be within the specified limits of $t_\alpha$ is $1 - \alpha$. This latter term is called the *level of confidence*. Note that $\alpha$ can be wholly to the right or to the left or divided between right and left. A table indicating the values of $t_\alpha$ for different degrees of freedom and values of $\alpha$ is given in Part 3.

**1–9–6 Comparison of Means of Two Samples**   It is often important to see whether or not the means of two samples are significantly different. Suppose, for example, we have two groups, A and B, of tensile test specimens, and that sample A was subjected to certain metallurgical treatment. If the mean of the tensile strength of sample A differs significantly from that of sample B, this leads to the conclusion that such a metallurgical treatment had an effect on the tensile strength. Usually the procedure followed is to determine $t$ for the two samples together using the following equation:

$$t = \frac{\bar{x}_1 - \bar{x}_2}{\sqrt{\left(\dfrac{S_1^2}{N_1}\right) + \left(\dfrac{S_2^2}{N_2}\right)}}$$

1–9

Next, the number of degrees of freedom (df) for the two samples together can be obtained from the following equation:

$$\gamma = \frac{\left[\left(\dfrac{S_1^2}{N_1}\right) + \left(\dfrac{S_2^2}{N_2}\right)\right]^2}{\dfrac{\left(\dfrac{S_1^2}{N_1}\right)^2}{N_1 - 1} + \dfrac{\left(\dfrac{S_2^2}{N_2}\right)^2}{N_2 - 1}}$$

1–10

$\gamma$ is rounded to the nearest integer and is used to obtain the interval $\pm t_{\alpha/2}$ (or $t_\alpha$) from Table 3–2 for any desired level of confidence $(1 - \alpha)$. If the value of $t$ obtained from Equation 1–9 does not fall within the interval $\pm t_{\alpha/2}$, it can be stated that the two means $\bar{x}_1$ and $\bar{x}_2$ are significantly different at that desired confidence level $(1 - \alpha)$.

# PART 2

## Laboratory Experiments

# EXPERIMENT 1  General Metallography

## OBJECTIVE

To enable the student to acquire familiarity with the techniques of preparing micro-sections for metallographic examinations and to become acquainted with the different pieces of equipment found in a typical metallurgical laboratory.

## BACKGROUND

Read the sections in this manual on the metallurgical microscope, preparation of specimens for metallographic examination, and etching of specimens.

## PROCEDURE

1. Follow the steps and directions explained earlier in this manual and prepare a few microsections for metallographic examination. Try to select your specimens from different ferrous and nonferrous alloys. Samples from low carbon steel, copper or brass, and aluminum or aluminum alloy would make a good combination.

2. Examine each specimen under the metallurgical microscope using different magnifications, such as 50×, 100×, and 400×. Make sure that the grain boundaries are clearly visible.

3. Store your specimen in the dessicator for the next experiment.

# EXPERIMENT 2  Determination of the Grain Size for a Microsection

## OBJECTIVE

To enable the student to become acquainted with photomicrography and make use of it in the determination of the grain size for a microsection.

## BACKGROUND

Read about the grain size number in the textbook used for the lectures. [Section 4–7 in Ref. 1]

## PROCEDURE

1.  Take a photomicrograph (using the metallurgical microscope) of one of the specimens you prepared in Experiment 1 and stored in the dessicator. Seek the assistance of the instructor when selecting the camera setting. It usually takes more than one attempt to obtain the optimal camera setting that yields clear photographs (called photomicrographs) in which the grain boundaries are evident and well defined. The magnification used should be higher than 100×.

2.  Using a sharp pencil, draw a circle or a rectangle of known area on the photomicrograph. Determine the nominal number of grains in that area, counting each whole grain as one and each boundary grain (i.e., the grains intercepted by the perimeter of the outlined area) as half.

3.  Calculate the number of grains per square inch at the magnification you used to obtain the photomicrograph. Modify the calculation to obtain the number of grains per square inch at 100× magnification, using the following equation:

$$n_{100} = n_M \cdot \left( \frac{M}{100} \right)^2$$

where $n_{100}$ is the number of grains per square inch at 100× magnification.

$n_M$ is the number of grains per square inch at the magnification $M$ of the photomicrograph.

4.  The ASTM grain size number $N$ (See ASTM E 112) is calculated from the following equation:

$$N = \frac{In(n)}{In(2)} + 1$$

5. A simpler method for determining the grain size number is the ASTM comparative method. It involves comparing the image of the structures (or a photomicrograph) at a magnification of 100× with a series of graded standard grain-size templates. Sometimes a grain-size measuring eyepiece is employed and it is basically based on the same idea of the comparative method.

# EXPERIMENT 3 Determination of the Lead-Tin Phase Diagram by Thermal Analysis

## OBJECTIVE

To construct the phase diagram for a simple binary system by thermal analysis. The Pb-Sn eutectic system was chosen because of its simplicity and the relatively low melting temperatures of these two metals.

## BACKGROUND

Read about the phase diagrams, the eutectic system, and the cooling curves for different molten alloys. [Sections 10–4 and 10–5 in Ref. 1]

## EQUIPMENT

Several Pyrex glass test tubes (or crucibles), Bunsen burners, wax, and a 3-channel continuous graph recorder (6-channel recorder would be better, of course).

## MATERIALS

Pure lead in the form of sheets or granules, pure tin in the form of sheets or granules.

## PROCEDURE

1. The test tubes are to be filled with mixtures of lead and tin in differing proportions. The mass of each charge should be about 50 g and the tin content should range between 0% (i.e., the whole charge is lead) and 100% (i.e., the whole charge is tin) in increments of 10% or 15%. If the lead and tin are in the form of sheets, they must be cut into small pieces, and folded not only to fit in the tubes, but also to occupy the smallest possible volume.

2. Using the Bunsen burners, heat the test tubes to bring each charge to its molten state, stirring occasionally with a glass rod to ensure the homogeneity of the molten alloy. At this point, add wax to each test tube to form a molten layer floating on the surface of the alloy to prevent excessive oxidation.

3. Immerse a thermocouple in each molten alloy, shut off the Bunsen burners to allow the molten alloys to cool down, and finally activate your graph recorder to obtain the cooling curve (i.e., temperature versus time) for each alloy on graph

paper. Usually, each curve will have a different color, and you should write the composition of the alloy next to its cooling curve to eliminate any confusion.

4. Shut off the graph recorder, then reheat the alloys to the molten state so that you can withdraw the thermocouples. Finally, wipe each thermocouple with a piece of cloth and shut off the Bunsen burners.

5. Obtain the phase transition temperatures for each alloy. Those are the points of inflection on the cooling curve, and they signify a phase change in the solidifying alloy.

6. Construct the phase diagram by drawing a light vertical line to represent each alloy, then plotting the phase transition temperatures on each of those lines, and finally connecting those points to indicate the liquidus and solidus lines.

7. If possible, obtain a sample from each alloy for metallographic examination.

# EXPERIMENT 4  Recrystallization of Brass

## OBJECTIVE

To determine the recrystallization temperature of cold-worked brass and to examine the effects of recrystallization on the mechanical properties (hardness) and the microstructure of the brass specimen.

## BACKGROUND

Study the theory and experimental principles of hardness testing as well as the method of operation of the Rockwell hardness tester in this manual. Also, review mechanical testing and properties, cold working, and recrystallization from your materials science textbook. [Sections 6–1 to 6–8 and 7–1 to 7–10 in Ref. 1]

## MATERIALS

Cold-rolled strip of brass (70/30 or 62/38).

## PROCEDURE

1. Cut a series of 6 or 7 specimens 12 mm × 12mm (0.5 in. × 0.5 in.) from the cold-rolled brass stock and remove burrs from the samples by filing or grinding.

2. Take a few readings of the hardness for each specimen and calculate the average Rockwell hardness for each specimen.

3. Put one sample aside and anneal the rest at different temperatures that completely cover the recrystallization temperature range for brass 150–700°C (300–1300°F). This is achieved by preheating a group of small laboratory resistance furnaces to different temperatures selected for developing recovery, recrystallization, and grain growth conditions, then placing the samples in the furnaces for at least half an hour.

4. Determine the mean Rockwell hardness for each of the annealed specimens.

5. Use your experimental data to plot a graph indicating Rockwell hardness versus the annealing temperature. The temperature at which there is a significant drop in hardness is the recrystallization temperature.

6. Prepare two microsections, one from the annealed sample that showed a significant drop in hardness and the other from the as-received (cold-worked) sample

that was spared from annealing. Make sure that the surface to be examined under the microscope is parallel to the direction of rolling. (You can also plot the grain size number versus the annealing temperature.)

7.  Subject these two microsections to metallographic examination and take a photo-micrograph for each.

## Review Questions on Recrystallization of Brass

1.  What is the recrystallization temperature?

2.  What are the main factors affecting the recrystallization temperature of a pure metal?

3.  What is the main difference between recovery and recrystallization?

4.  Why is recrystallization accompanied by an appreciable drop in the hardness of a specimen?

5.  In what way does the microstructure of a specimen, after recrystallization, differ from that of its as-cold-worked condition?

6.  What effect will recrystallization have on the mechanical properties of a metallic specimen?

7.  If a specimen is left at an elevated temperature for a long time after recrystallization takes place, what will happen?

8.  What are the stages of the process of recrystallization?

9.  What is the driving force for recrystallization?

10. What is the difference between hot working and cold working of metals?

11. List two advantages that hot working of metals has over cold working.

12. List two advantages that cold working of metals has over hot working.

13. What is the main driving force for grain growth?

14. Is it better to use fine-grain metals or coarse-grain metals for load-carrying components at room temperature? Why?

# EXPERIMENT 5  Precipitation Hardening of Aluminum Alloys

## OBJECTIVE

To examine the effects of aging time and temperature on the mechanical properties (hardness) and the microstructure of a common heat-treatable aluminum alloy (2024).

## BACKGROUND

Knowledge of the operation principles of the Rockwell hardness tester. Review the sections on precipitation hardening in nonferrous metals, heat-treatable aluminum alloys, and the Al-Cu phase diagram in your materials science textbook. [Sections 11–4 to 11–7 and 13–2 in Ref. 1]

## MATERIALS

Strip of aluminum alloy 2024-T3 (Al-Cu alloy containing magnesium, manganese, silicon, and chromium in addition to copper). In its fully heat-treated condition (T6), it is the most commonly used structural alloy in subsonic aircrafts. T3 indicates that the alloy was subjected to solution treatment followed by strain hardening and natural aging.

## PROCEDURE

1. Cut four specimens 12 mm × 12 mm (0.5 in. by 0.5 in.) of the strip and remove the cutting burrs.

2. Stamp identification marks on the specimens in such a manner that they will not interfere with or affect the hardness measurements.

3. Determine the average Rockwell hardness for each specimen.

4. Place each of the first three specimens in a furnace at a temperature of about 510–540°C (950–1000°F) for 45 to 60 minutes (this is called solution heat treatment). Keep the fourth specimen for metallographic examination in its as-received condition.

5. Remove the specimens from the three furnaces and quench them in water as quickly as possible.

6. Determine the average Rockwell hardness for each of the quenched specimens.

7. Put one of the quenched specimens in the freezer to retard the aging process and determine its Rockwell hardness after two weeks.

8. Keep another quenched specimen at room temperature (this is called natural aging or natural age hardening). Determine its Rockwell hardness every hour for the first six hours, then every other day for two weeks. Plot your data in the form of hardness versus logarithm of aging time.

9. Place the third quenched specimen in the furnace for about one hour at a temperature of 190°C (375°F) (this is called artificial aging). Next, cool the specimen slowly to room temperature and measure its Rockwell hardness.

10. Compare the final Rockwell hardness for the four specimens.

11. Subject all four specimens to metallographic examination and try to make a correlation between the microstructure of each specimen and its hardness.

## Review Questions on Precipitation Hardening of Aluminum Alloys

1. Briefly explain the purpose of each of the steps involved in the age-hardening process.

2. Why is it important that the specimen (which has just been subjected to solution heat treatment) be taken very quickly from the furnace and dropped into the quench bath?

3. Is it appropriate to place your specimen close to the door of the furnace during solution heat treatment?

4. What results would be expected if you solution heat treat your specimen of 2024-T3 aluminum alloy at: (a) 650°C (1200°F); (b) 550°C (1025°F); (c) 425°C (800°F)?

5. What is the role of solid-state diffusion in the precipitation hardening process?

6. What is the difference between coherent and noncoherent precipitates?

7. Using your results as a reference, explain the meaning of the term "incubation period."

8. Using your results as a reference, can you predict how the hardness versus aging time curve would change for different aging temperatures?

9. Why are the rivets made of heat-treatable aluminum alloys stored in dry ice before being riveted into place?

10.  Does the cold forming of an aluminum alloy, which has been solution heat treated and quenched, have any effect on the age-hardening process? Why?

11.  During artificial aging, if the specimen is left for a longer time (over-aged), how would this affect its hardness and its microstructure?

# EXPERIMENT 6  Annealing and Normalizing of Steel

## OBJECTIVE

To study the effect of the cooling rate and the amount of alloying elements on the micro-structure and the mechanical properties (Rockwell hardness) of medium carbon steel.

## BACKGROUND

Study the method of operation of the Rockwell hardness tester used in your laboratory. Also, carefully review the iron-carbon phase diagram, isothermal transformation diagrams, and the annealing and normalizing operations in your materials science textbook. [Sections 12–1 and 12–4 to 12–6 Ref. 1]

## MATERIALS

A stock (bar or hexagonal) of plain carbon steel AISI 1045 and a stock of low alloy steel AISI 4140. The approximate compositions of these two kinds of steel are given in your textbook. They both have the same carbon content, about 0.4%, but the second contains other alloying elements such as molybdenum.

## PROCEDURE

1. Cut two small pieces 12 mm × 12 mm (about 0.5 in. × 0.5 in.) from the plain carbon steel stock, and two similar pieces from the low alloy steel stock.

2. Remove burrs from the samples by filing or grinding and stamp an identification code on each sample.

3. Place each sample in a small resistance furnace which is set at the proper autenitizing temperature, about 845°C (1550°F). The temperature of the furnace will first drop then recover again to the set temperature. Keep each specimen in the furnace half an hour after the temperature returns to its initial setting.

4. Take one plain carbon steel specimen and one low alloy steel specimen out of the furnaces and let them cool down to room temperature (place them on a refractory tile or a brick in still air). As explained in your textbook, this is called a normalizing operation.

5. Furnace cool the other two specimens (one plain carbon steel and one low alloy steel). This is achieved by shutting off the furnace while keeping its door closed and the specimen inside, thus ensuring a very low cooling rate, which can actu-

ally be determined. This operation is called annealing. Remember, this takes a very long time so you can let the furnace cool down overnight.

6. Measure the Rockwell hardness of each specimen. If possible, use scale B or even scale A, so that the Rockwell hardness of the two annealed and two normalized specimens would fall on the same scale, thus facilitating comparison.

7. Prepare a microsection from each specimen for metallographic examination, and take a photomicrograph for each specimen.

8. Compare the Rockwell hardness of the different specimens as well as their photomicrographs and try to draw a correlation between the microstructure and the Rockwell hardness.

## Review Questions on Annealing and Normalizing of Steel

1. What is the effect of increasing the cooling rate (from furnace cooling to air cooling) on the fineness of pearlite formed and on the amounts of the different phases for the AISI 1045 specimens and AISI 4140 specimens?

2. What is the effect of increasing the cooling rate (from furnace cooling to air cooling) on the hardness of the AISI 1045 specimens and the AISI 4140 specimens?

3. Explain why it is very difficult to determine the carbon content of a normalized steel specimen from its microstructure only.

4. Estimate the amount of pearlite in each specimen and compare these amounts with the equilibrium values obtained from the iron-carbon phase diagram.

5. Mention some factors that may affect the actual cooling rate of an air-cooled specimen.

6. What effects do the alloying additives have on the microstructure and mechanical properties of air-cooled (normalized) steel specimens?

7. What effects do the alloying additives have on the microstructure and mechanical properties of furnace-cooled (annealed) steel specimens?

# EXPERIMENT 7 Quenching and Tempering of Steel

## OBJECTIVE

To study the effects of quenching and tempering of plain carbon eutectoid steel on its microstructure and its mechanical properties. The experiment can also be expanded to investigate the effects of the carbon content on the mechanical properties and microstructure of plain carbon steel.

## BACKGROUND

Knowledge of the operation principles of a Rockwell hardness tester. Review the iron-carbon phase diagram, the isothermal transformation diagram, and the heat treatment operations in your materials science textbook. [Sections 11–9 to 12–5 in Ref. 1]

## MATERIALS

A stock of AISI 1080 steel (AISI 1095 can be used if AISI 1080 is not available).

## PROCEDURE

1. Cut six specimens 12 mm × 12 mm (0.5 in. × 0.5 in.) from the stock and remove burrs by filing and/or grinding.

2. Stamp the specimens for identification and determine the average Rockwell hardness for each. Be careful not to take hardness measurements near any stamped area.

3. Set aside one of the specimens for later metallographic examination, and put each of the others in a furnace with the temperature set at 830°C (1530°F), about 100°C (180°F) above $AC_3$ as indicated by the phase diagram. The temperature of the furnace will drop slightly; as soon as it recovers, start measuring 45 minutes.

4. After the autenitizing time elapses, take each specimen out of the furnace quickly and drop it in an appropriate quenching tank containing two to four liters of water (about one-half to one gallon of water for each specimen).

5. Take the specimens out of the water and measure the Rockwell hardness of each.

6. Set aside one specimen for later metallographic examination and place the other

four specimens into four resistance furnaces set at temperatures of 205°C (400°F), 315°C (600°F), 425°C (800°F), and 540°C (1000°F), respectively. This process is called tempering.

7.  After tempering the specimens for 45 minutes, take them out and allow them to cool down.

8.  Measure the Rockwell hardness for each of these tempered specimens and plot a graph indicating hardness versus tempering temperature. Using horizontal lines of different colors, also show the original hardness of the as-received specimen as well as the hardness after quenching.

9.  Prepare microsections for all the as-received, quenched, and tempered specimens, and subject them to metallographic examination.

10. Try to provide an explanation for the mechanical properties (hardness) obtained using the observed microstructure.

## Review Questions on Quenching and Tempering of Steel

1.  How do you describe the microstructure of the as-received specimen?

2.  Can you predict the mechanical properties of that specimen from the observed microstructure?

3.  What happens to a specimen when it is placed in a furnace at 830°C (1530°F)?

4.  Why is the hardness of a quenched specimen so high? Is it related to the microstructure? If yes, in what way?

5.  What happens to a quenched specimen during the tempering operations? What causes that change?

6.  What are the effects of the tempering temperature on the mechanical properties (hardness, tensile strength, and ductility) of the tempered specimens?

7.  Describe the microstructure of each of the tempered specimens. What is the effect of the tempering temperature on the microstructure?

8.  How do you explain the hardness versus tempering temperature graph in light of your observations of the microstructure of the tempered specimens?

9.  What is the industrial significance of the hardness-tempering temperature plot (i.e., how can we make use of it in industry)?

10. Can you anticipate the results if the same experiment were performed on AISI 1020 steel? Explain.

11. Can you anticipate the results if the same experiment were performed on AISI 1095 steel? Explain.

12. What is the effect of carbon content on the mechanical properties of quenched and of tempered steels?

# EXPERIMENT 8  Tensile Testing of Metals

## OBJECTIVE

To obtain the engineering stress–engineering strain curve for soft plain carbon steel (low carbon, i.e., AISI 1012 or 1020) as well as for a copper-base alloy, and to determine the important mechanical properties of these two alloys from the obtained curves. Also, to try to understand the mechanics of deformation in each case. In addition, the experiment can be expanded to investigate the effects of the strain rate on the mechanical properties of the alloys.

## BACKGROUND

Full knowledge of the operation of the universal tensile testing machine and an appropriate type of strain-indicating device. Also, familiarity with safety precautions to stop the machine promptly in an emergency situation. Review tensile testing of materials, the engineering stress-strain curve, and the mechanical properties of materials in your materials science textbook. [Sections 6–1 to 6–6 in Ref. 1]

## EQUIPMENT

A universal tensile testing machine similar to that shown in Figure 2–1 together with an appropriate strain-indicating device, point micrometer and vernier calliper, hardness testing machine.

## MATERIALS

A standard tensile test specimen 12.83 mm in diameter (0.505 in.) of each material (plain carbon steel and copper-base alloy). A drawing of that specimen is shown in Part 3 (ASTM Designation: E 8).

## PROCEDURE

1. Use a V block or an appropriate fixture to place two marks 50 mm (2.0 in.) apart (using a punch, a pencil, or a marker) on each test specimen. This will be the gage length. Divide it into eight equal parts (each 6.35 mm or 0.25 in.) and place a mark (with pencil or marker). Measure the diameter of the test specimen and record it.

2. Secure the ends of the test specimen in the grips of the universal testing machine.

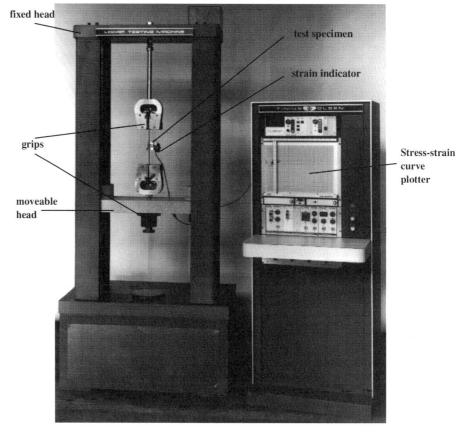

fixed head

test specimen

strain indicator

grips

Stress-strain
curve
plotter

moveable
head

**FIGURE 2–1** A universal tensile testing machine. (*Courtesy of Tinius Olsen Testing Machine Co., Inc., Willowgrove, PA.*)

**3.** Mount or attach the strain-indicating device.

**4.** Apply the load very slowly to insure that it is quasi-static (use a very low strain rate of about 3 mm/min or ⅛ in./min). Take strain readings at reasonably spaced levels of load (i.e., the increment of load should be about one ton). Be very careful to record the upper and lower yield loads for the steel specimen as well as the corresponding strain readings.

**5.** After the yield point is reached and before necking of the specimen occurs, remove the strain-indicating device (to avoid breaking it). Proceed with the test to obtain the maximum load and the fracture load.

**6.** Remove the fractured test specimen from the grips and measure the final gage length as well as the final length between each two successive marks on the specimen (the original length of this distance is 6.35 mm or 0.25 in.). Also, measure the diameter of the specimen at the fracture (the neck).

7. Use a V block or a fixture to take Rockwell hardness readings along the length of the specimen, at midpoint between each two successive marks.

8. Plot the stress-strain curve using an appropriate scale to enable calculating the modulus of elasticity. Determine the yield point (upper and lower points for the plain carbon steel specimen and 0.5% offset yield point for the copper alloy specimen), the ultimate strength, the percentage elongation, and the reduction in area for each of the two specimens.

9. Plot the elongation percent for each of the (6.35 mm or 0.25 in.) divisions versus distance along the gage length for each specimen. Also, plot the Rockwell hardness versus distance along the gage length. Try to correlate those two curves.

10. Repeat the experiment using the maximum strain rate that can be provided by the testing machine (set the speed of the head to maximum). Compare the results you obtain with the previous ones.

## Review Questions on Tensile Testing of Metals

1. What divides the engineering stress-strain curve into two regions, namely, the elastic and plastic regions?

2. What is the difference between the behavior of the material in the elastic and plastic regions of the engineering stress-strain curve?

3. What measurement should you have taken in order to be able to plot the true stress–true strain curve?

4. Why does low carbon steel have clear upper and lower yield points? Why doesn't the copper have the same? Explain the differences using your knowledge of the alloying and dislocation theories.

5. Did you observe parallel lines inclined about 45° with the horizontal on the surface of the steel specimen, and in the area where necking later took place? What are those lines called? What do they reveal about the mechanism of deformation? Apply Schmid's law.

6. Could you observe any similar lines on the copper alloy specimen? What is the mechanism of deformation in this latter case?

7. Obtain the area under the load-extension curve (i.e., energy) for each specimen and divide it by the volume between the gage length in order to obtain the modulus of toughness. Compare the value of plain carbon steel and that of copper. Can you draw any conclusion?

8. Looking at the curve indicating the distribution of elongation along the gage length, where did the maximum localized elongation take place?

9. How do you interpret the shape of the above-mentioned curve?

10. Is there any similarity between the distribution of the localized elongation along the gage length and that of the hardness measurements? Explain why.

11. What effect does the high strain rate have on the mechanical properties mentioned below?
    - Modulus of elasticity
    - Yield stress
    - Ultimate tensile strength
    - Elongation percent
    - Modulus of toughness

    How do you explain the reasons for those effects?

12. Did the high strain rate have an effect on the mode of failure of any of the specimens? Why?

# EXPERIMENT 9  Tensile Testing of Polymers

## OBJECTIVE

To understand the effect of the molecular structure of polymers on their behavior when subjected to mechanical loading, and to compare their mechanical properties with those of the plain-carbon steel and copper-base alloy previously obtained.

## BACKGROUND

Knowledge of the operation of the universal tensile testing machine. Review tensile testing and mechanical properties of materials, as well as the chapter covering polymers in your materials science textbook. [Sections 15–7 and 15–10 and Chapter 6 in Ref. 1, and ASTM D638-86]

## EQUIPMENT

A universal tensile testing machine capable of directly plotting the load-extension curve for the tested specimen (a Hounsfield tensometer is adequate for this experiment).

## MATERIALS

A standard flat test specimen of high density polyethylene, and another one made of a thermosetting material like epoxy or phenolics, etc. These specimens can be obtained by machining polymeric sheets or can be cut from the sheet by a simple blanking die. Specimens are also readily available from suppliers of scientific equipment.

## PROCEDURE

1.  Mount the high-density polyethylene specimen on the tensile testing machine and obtain the engineering stress-strain curve as previously explained in Experiment 8. The loading increment should, however, be in the order of 1000–1500 $N$ (200–300 pounds) only.

2.  Repeat the same procedure for the thermosetting specimen.

3.  Compare the above-mentioned two curves with those of the plain-carbon steel and of the copper-base alloy. Compare the mechanical properties of polymers with those of metals.

4.  Perform the tension test again on two polymer specimens similar to the previous

ones, but using a higher strain rate. Study the effects of higher strain rate on the yield strength and ultimate tensile strength of the polymers.

## Review Questions

1. Did fracture occur at the neck for the high-density polyethylene specimen?

2. After necking of the polyethylene specimen, did further deformation occur at the neck area?

3. What happened to the color and transparency of the polyethylene specimen at the neck area?

4. Were similar phenomena observed when testing the thermosetting specimen?

5. Are the modes of failure of the thermoplastic and thermosetting specimens similar?

6. What are the effects of high strain rate on the failure mode and on the mechanical properties of each of the thermoplastic and thermosetting specimens?

7. Does the failure mode of polymers differ from that of metals? Why?

8. What is the elastic deformation mechanism for high-density polyethylene? Is it different from that of metals? In what way?

9. What is the plastic deformation mechanism for high-density polyethylene? How does it affect the shape of the engineering stress-strain curve?

10. Discuss how the microstructures of polymers and metals make their behavior under mechanical loading different.

# EXPERIMENT 10  Impact Testing of Polymers

## OBJECTIVE

To study the effect of temperature on the behavior of different polymers under impact loading. Also, to determine the glass transition temperature of one or more of those polymers.

## BACKGROUND

Knowledge of the operation of the impact testing machine. Review impact testing and the mechanical properties of polymers. [Chapter 6, Section 15–7 in Ref. 1]

## EQUIPMENT

Impact testing machine, either the conventional 163 J (120 ft.lb.) machine or a smaller one suitable for polymers testing; thermocouples with digital readout; acetone-solid $CO_2$, powdered dry ice, and liquid-nitrogen baths for lowering the temperatures of the specimens.

## MATERIALS

If the conventional impact testing machine is used, it is advisable to use charpy unnotched specimens of the different polymers, each having a cross section of 10 mm × 10 mm (3/8 in. × 3/8 in.). Polymers to be tested can include high-density polyethylene, polystyrene, nylon, etc.

## PROCEDURE

1. A set of identical test specimens of the same polymer are to be subjected to impact testing at different temperatures. The first specimen is tested at room temperature, and the energy consumed to break it is determined in the usual manner.

2. Place the second specimen in the acetone-solid $CO_2$ bath for a few minutes, then transfer it quickly to the testing machine and perform the impact test. Use a thermocouple to measure the temperature of the specimen just before the pendulum of the machine hits it. Record both the temperature and the energy used to break the specimen.

3. Repeat the previous step a few times, changing the soaking period of the speci-

men in the bath in each case in order to change the temperature of the specimen during impact.

4. Repeat the above-mentioned procedures but use the liquid nitrogen bath so that the impact test can be performed on specimens at lower temperatures.

5. Plot the impact strength of a specimen versus its temperature during the test.

6. The glass transition temperature for that polymer is the temperature at which there is a sudden drop in its impact strength.

7. Use the same procedure to obtain the glass transition temperature for each of the other polymers.

## Review Questions on Impact Testing of Polymers

1. How do you compare the impact strengths of polymers with those of metals? Explain the reason for any differences.

2. Would the presence of a deep scratch on the surface of a polymeric specimen affect its impact strength? If the answer is yes, what do we call this characteristic property?

3. Would a similar scratch have a similar effect on the impact strength of a metal specimen?

4. What is the percentage drop in impact strength at the glass transition temperature for each of the polymers tested?

5. Is this phenomenon observed in all metals? Explain.

# EXPERIMENT 11  Slip Casting of Ceramics

## OBJECTIVE

To gain hands-on experience in the processing of ceramics by slip casting to produce a part in its final shape; also, to identify and try to understand the parameters affecting the operations of drying and sintering of the ceramic product.

## BACKGROUND

Review sections on processing of ceramics and applications and properties of ceramics in your materials science textbook. [Sections 14–8 and 14–9 in Ref. 1]

## EQUIPMENT

A furnace suitable for sintering the green ceramic part, pH value meter and viscosity meter, point micrometer, and vernier caliper.

## MATERIALS

A split mold of the desired part made of plaster of Paris and a ceramic slip. Both can be obtained from a ceramic supply house, or they can be prepared by the student. In the latter case, further reading specifically about the slip casting method is required. Appropriate references are provided in the bibliography of this manual.

## PROCEDURE

1.  The slip, which is actually a suspension of fine ceramic particles in a viscous liquid, is blended so that it would form a thick laminar stream when poured. Make sure that the viscosity of the slip matches the value recommended by the supplier. You can always control the viscosity as needed by adjusting the pH valve of the slip using drops of an acid (and an alkaline).

2.  After clamping the two halves of the mold, the slip is poured slowly and steadily into the mold, so that no air bubbles will be formed.

3.  Wait till an adhering solid layer of about 6 mm (0.25 in.) builds up, then invert the mold slowly to pour off the remaining liquid slip.

4.  Allow that cast layer to dry in place for at least 20 minutes, then open the mold and remove the part (assuming your part is relatively small).

5.  Allow the part to dry for several days, while carefully controlling the temperature and humidity of the atmosphere surrounding it. The recommended environment involves a warm and low-humidity atmosphere. Before firing, measure the dimension of the green part.

6.  This step, which is called biscuit or bisque firing, is, in fact, a sintering operation, which results in an increase in the rigidity and strength of the ceramic part. It is a good approach to sinter similar green parts at a variety of temperatures and study the effects of the firing temperature on the physical and mechanical properties.

7.  Study the shrinkage characteristics of the part at various firing temperatures by measuring the dimensions before and after firing and calculating the percentage of shrinkage across the diameter and the height.

## Review Questions on Slip Casting of Ceramics

1.  Is it necessary that the mold be porous and absorbant? Why?

2.  Why should the temperature and humidity be carefully controlled during the drying operation?

3.  If the humidity is too low, what would happen to the ceramic part?

4.  If the humidity is too high, what would happen to the ceramic part?

5.  Why does the strength of a ceramic part increase as a result of firing?

6.  What effects do the firing temperature have on the shrinkage during firing?

7.  Plot the shrinkage across the diameter versus the firing temperature.

8.  Plot the shrinkage across the height versus the firing temperature.

9.  Why does shrinkage occur during firing?

# EXPERIMENT 12  Diametral Compression Test of Ceramics

## OBJECTIVE

To determine a reliable value for the tensile strength of the ceramic material given and to study the mode of failure under tension.

## BACKGROUND

Review the chapter on ceramics in your materials science textbook [Chapter 14 in Ref. 1]. Also review Mohr's Circle for stress in any mechanics of materials book.

## EQUIPMENT

A universal tension/compression testing machine with compression test platens, and a micrometer or a vernier caliper.

## MATERIALS

A group of flat disks, all having the same dimensions and made of the ceramic material for which the tensile strength is to be determined. A recommended value for the diameter is 50–75 mm (2–3 in.) and the recommended thickness is 12–25 mm (0.5 to 1.0 in.).

## PROCEDURE

This test, which is also known as the Brazilian test or the splitting tension test, involves compressing a flat disk diametrically between two flat platens. Fracture occurs along the loaded diameter as a result of tensile stress initiated across this diameter. That stress acts uniformly across the diameter and is given by the following equation:

$$\sigma = \frac{2P}{Dt\,\pi}$$

where $P$ = applied load

$\quad\quad D$ = specimen diameter

$\quad\quad t$ = specimen thickness

1.  Using the vernier caliper (or the micrometer), measure the diameter and the thickness of each of the ceramic disks to be tested.

2.  Place the first disk between the two parallel compression platens and apply the load slowly until the disk is fractured. Record the fracture load. The parallelism

of the platens and the alignment of the loaded diameter with the axis of the machine are critical factors.

3.  Determine the tensile strength for the ceramic material using the above-mentioned equation.

4.  Repeat the first three steps for the other disks.

5.  Calculate the mean value of the tensile strength for the group and also the standard deviation.

6.  Using the statistic $t$, determine the mean value of the tensile strength for an infinite number of disks at a confidence level of 95%.

## Review Questions on Diametral Compression of Ceramics

1.  Examine the fractured surface and comment on it.

2.  List some advantages and disadvantages of this test.

3.  What are the sources of error in the experiments?

4.  Would you recommend that this test be carried out on low carbon steel? If not, explain why.

5.  In your opinion, would any variation in the thickness of the disk have an effect on the results obtained? Support your answer with evidence.

6.  What effect would the mechanical properties of the platens have on the results? Explain why.

7.  Why wouldn't the material fail in compression rather than in tension?

# EXPERIMENT 13  Flexure Test of Ceramics

## OBJECTIVE

To determine the flexural strength of a given ceramic material (or glass) and to study the effect of the size of the specimen tested on the value obtained.

## BACKGROUND

Review the chapter on ceramics in your materials science textbook [Chapter 14 in Ref. 1]. Also review bending of beams in any mechanics of materials book and see ASTM Standards, Designation: C158-43.

## EQUIPMENT

A universal testing machine with a three-point flexural bending test assembly; micrometer or a vernier caliper.

## MATERIALS

Two groups of rectangular cross-section bars, all having the same thickness of about 3–4 mm (1/8–1/6 in.). The bars of the first group A would all have a width of one inch while each of those of the second group B should be 2 inches (possibly 3 inches) wide. They can be all of a given ceramic material or they can be glass (which is less expensive). Make sure that the sharp edges of the specimens on the tension side are chamfered or rounded before a test is performed, in order to minimize erratic error.

## PROCEDURE

1. Using the vernier caliper (or micrometer), measure the width and thickness of each specimen.

2. Place the specimen in the flexure test assembly. Apply load and increase it uniformly until the specimen fractures and measure the load at which the specimen breaks.

3. Calculate the flexural strength, also known as the modulus of rupture, from the following equation:

$$\sigma = \frac{Mc}{I}$$

where $\sigma$ is the flexure strength

$M$ is the bending moment resulting from the fracture load

$c$ is the half thickness of the specimen

$I$ is the moment of inertia of the cross section of the specimen around the neutral axis.

4. Repeat the previous steps for all the specimens of group A and of group B.

5. Calculate the mean value of the flexural strength and the standard deviation for each group.

6. Apply the significant test ($t$ test) to see whether or not there is a significant difference between the mean value of flexural strength for group A and that for group B.

**Review Questions on Flexure Test of Ceramics**

1. What are the causes of the scatter in the values of the flexural strength for each group?

2. List some advantages and some disadvantages of this test.

3. Is it better to have a three-point test or a four-point test? Support your answer with evidence.

4. Is the obtained modulus of rupture an indication of the tensile strength or of the compressive strength? Why?

5. Is there a significant difference between the mean value of the flexure strength for group A and that for group B, for a level of confidence of 95%? If there is, how can you explain the reason?

6. Examine the fractured surface of each of the specimens. How can you comment on it?

7. If the test is carried out on wrought iron, would the shape of the fractured surface be different?

# EXPERIMENT 14  Tensile Testing of Composites

## OBJECTIVE

To study the behavior of fiber-reinforced polymeric composite under uniaxial tensile loading. Also, to try to understand the mechanics of deformation and the mode of failure.

## BACKGROUND

Full knowledge of the operation of the universal tensile testing machine. Review the chapter on composites in your materials science textbook. [Chapter 16 in Ref. 1]

## EQUIPMENT

A universal tensile testing machine; vernier caliper or micrometer.

## MATERIALS

A standard flat tensile test specimen of FRP, glass-filled polyester, or nylon 66 are suitable material; specimens are readily available from suppliers of scientific equipment.

## PROCEDURE

1. Secure the ends of the test specimen in the grips of the universal testing machine.

2. Apply the load very slowly and obtain the engineering stress-strain curve as previously explained.

3. Examine the fracture surface visually with a magnifying glass.

4. Determine the different mechanical properties for the composite specimens.

### Review Questions on Tensile Testing of Composites

1. How do you compare the engineering stress-strain curve for a composite with that for a pure (non-reinforced) polymer?

2. Show the difference in appearance between the above-mentioned two curves and

explain the reasons for these differences, referring to the mechanics of deformation and mode of failure.

3. How do the mechanical properties of a composite compare with those of a polymer?

4. How do the mechanical properties of a composite compare with those of low carbon steel? How would the result differ when you take weight into account?

5. Did failure occur as a result of matrix rupture, breakage of fibers, or pulling of fibers out of the matrix?

6. How would an increase in the percentage of fibers affect the shape of the stress-strain curve and the mode of failure?

# EXPERIMENT 15  Temperature Effect on Viscoelasticity of Polymers

## OBJECTIVE

To study the effect of time and temperature on the viscoelastic behavior of a given polymeric material by taking isothermal stress relaxation measurements over a range of temperatures.

## BACKGROUND

Full knowledge of the operation of the universal tensile testing machine. Review the chapter on polymers in your materials science textbook. [Chapter 15 in Ref. 1]

## EQUIPMENT

A universal tensile testing machine fitted with a temperature chamber; vernier caliper or micrometer.

## MATERIALS

A group of identical standard flat tensile test specimens of the given polymer (high-density polyethylene is suitable).

## PROCEDURE

1. Secure the ends of the test specimen in the grips of the universal testing machine.

2. Apply a load at room temperature to generate a stress just below the yield stress of the material (you can make use of the results of Experiment 9).

3. Start measuring the time and the corresponding load until the rupture occurs. Using the dimensions of the specimen, you can obtain the applied stress-time relationship.

4. On a log-log scale, plot the above-mentioned relationship.

5. Repeat steps 1 to 4, but increase the temperature by 20°C (36°F) and reduce the initial load by about 50%.

6. Repeat the procedure again several times, increasing the temperature each time

by 20°C (36°F) and reducing the initial load by 50% of the one used in the pre-ceding experiment.

## Review Questions on Temperature Effect on Viscoelasticity of Polymers

1. What effect does temperature have on the rupture stress?

2. What effect does temperature have on the rupture strain?

3. Does temperature have any effect on the deformation behavior and failure mode of a polymer? Explain how.

4. Did you notice any change in the nature of the fractured surface when increasing the temperature?

5. If you describe the behavior of the polymer at each temperature by relating it to a known engineering material (e.g., glass, rubber, etc.), describe the behavior of the polymer in the following temperature ranges:

   a. room temperature to 60°C

   b.  60°C to 120°C

   c.  120°C to 180°C.

# PART 3

## Reference Material

**TABLE 3–1**   Standard Normal Distribution Areas.

| z | .00 | .01 | .02 | .03 | .04 | .05 | .06 | .07 | .08 | .09 |
|---|-----|-----|-----|-----|-----|-----|-----|-----|-----|-----|
| | | | | | **Second Decimal Place in z** | | | | | |
| 0.0 | .0000 | .0040 | .0080 | .0120 | .0160 | .0199 | .0239 | .0279 | .0319 | .0359 |
| 0.1 | .0398 | .0438 | .0478 | .0517 | .0557 | .0596 | .0636 | .0675 | .0714 | .0753 |
| 0.2 | .0793 | .0832 | .0871 | .0910 | .0948 | .0987 | .1026 | .1064 | .1103 | .1141 |
| 0.3 | .1179 | .1217 | .1255 | .1293 | .1331 | .1368 | .1406 | .1443 | .1480 | .1517 |
| 0.4 | .1554 | .1591 | .1628 | .1664 | .1700 | .1736 | .1772 | .1808 | .1844 | .1879 |
| 0.5 | .1915 | .1950 | .1985 | .2019 | .2054 | .2088 | .2123 | .2157 | .2190 | .2224 |
| 0.6 | .2257 | .2291 | .2324 | .2357 | .2389 | .2422 | .2454 | .2486 | .2518 | .2549 |
| 0.7 | .2580 | .2612 | .2642 | .2673 | .2704 | .2734 | .2764 | .2794 | .2823 | .2852 |
| 0.8 | .2881 | .2910 | .2939 | .2967 | .2995 | .3023 | .3051 | .3078 | .3106 | .3133 |
| 0.9 | .3159 | .3180 | .3212 | .3238 | .3264 | .3289 | .3315 | .3340 | .3365 | .3389 |
| 1.0 | .3413 | .3428 | .3461 | .3485 | .3508 | .3531 | .3554 | .3577 | .3599 | .3621 |
| 1.1 | .3643 | .3665 | .3686 | .3708 | .3729 | .3749 | .3770 | .3790 | .3810 | .3830 |
| 1.2 | .3849 | .3869 | .3888 | .3907 | .3925 | .3944 | .3962 | .3980 | .3997 | .4015 |
| 1.3 | .4032 | .4049 | .4066 | .4082 | .4099 | .4115 | .4131 | .4147 | .4162 | .4177 |
| 1.4 | .4192 | .4207 | .4222 | .4236 | .4251 | .4265 | .4279 | .4292 | .4306 | .4319 |
| 1.5 | .4332 | .4345 | .4357 | .4370 | .4382 | .4394 | .4406 | .4418 | .4429 | .4441 |
| 1.6 | .4452 | .4463 | .4474 | .4484 | .4495 | .4505 | .4515 | .4525 | .4535 | .4545 |
| 1.7 | .4554 | .4564 | .4573 | .4582 | .4591 | .4599 | .4608 | .4616 | .4625 | .4633 |
| 1.8 | .4641 | .4649 | .4656 | .4664 | .4671 | .4678 | .4686 | .4693 | .4699 | .4706 |
| 1.9 | .4713 | .4719 | .4726 | .4732 | .4738 | .4744 | .4750 | .4756 | .4761 | .4767 |
| 2.0 | .4772 | .4778 | .4783 | .4788 | .4793 | .4798 | .4803 | .4808 | .4812 | .4817 |
| 2.1 | .4821 | .4826 | .4830 | .4834 | .4838 | .4842 | .4846 | .4850 | .4854 | .4857 |
| 2.2 | .4861 | .4864 | .4868 | .4871 | .4875 | .4878 | .4881 | .4884 | .4887 | .4890 |
| 2.3 | .4893 | .4896 | .4898 | .4901 | .4904 | .4906 | .4909 | .4911 | .4913 | .4916 |
| 2.4 | .4918 | .4920 | .4922 | .4925 | .4927 | .4929 | .4931 | .4932 | .4934 | .4936 |
| 2.5 | .4938 | .4940 | .4941 | .4943 | .4945 | .4946 | .4948 | .4949 | .4951 | .4952 |
| 2.6 | .4953 | .4955 | .4956 | .4957 | .4959 | .4960 | .4961 | .4962 | .4963 | .4964 |
| 2.7 | .4965 | .4966 | .4967 | .4968 | .4969 | .4970 | .4971 | .4972 | .4973 | .4974 |
| 2.8 | .4974 | .4975 | .4976 | .4977 | .4977 | .4978 | .4979 | .4979 | .4980 | .4981 |
| 2.9 | .4981 | .4982 | .4982 | .4983 | .4984 | .4984 | .4985 | .4985 | .4986 | .4986 |
| 3.0 | .49865 | .4987 | .4987 | .4988 | .4988 | .4989 | .4989 | .4989 | .4990 | .4990 |
| 4.0 | .4999683 | | | | | | | | | |

Source: Neter, Wasserman, and Whitemore, *Fundamental Statistics for Business and Economics,* 4th ed. (Boston: Allyn and Bacon, 1972).

**TABLE 3–2**   The t-Distribution.

| degrees of freedom | $\alpha \longrightarrow$ 0.10 | .05 | .025 | .01 | .005 |
|---|---|---|---|---|---|
| 1 | 3.078 | 6.314 | 12.706 | 31.821 | 63.657 |
| 2 | 1.886 | 2.920 | 4.303 | 6.965 | 9.925 |
| 3 | 1.638 | 2.353 | 3.182 | 4.541 | 5.841 |
| 4 | 1.533 | 2.132 | 2.776 | 3.747 | 4.604 |
| 5 | 1.476 | 2.015 | 2.571 | 3.365 | 4.032 |
| 6 | 1.440 | 1.943 | 2.447 | 3.143 | 3.707 |
| 7 | 1.415 | 1.895 | 2.365 | 2.998 | 3.499 |
| 8 | 1.397 | 1.860 | 2.306 | 2.896 | 3.355 |
| 9 | 1.383 | 1.833 | 2.262 | 2.821 | 3.250 |
| 10 | 1.372 | 1.812 | 2.228 | 2.764 | 3.169 |
| 11 | 1.363 | 1.796 | 2.201 | 2.718 | 3.106 |
| 12 | 1.356 | 1.782 | 2.179 | 2.681 | 3.055 |
| 13 | 1.350 | 1.771 | 2.160 | 2.650 | 3.012 |
| 14 | 1.345 | 1.761 | 2.145 | 2.624 | 2.977 |
| 15 | 1.341 | 1.753 | 2.131 | 2.602 | 2.947 |
| 16 | 1.337 | 1.746 | 2.120 | 2.583 | 2.921 |
| 17 | 1.333 | 1.740 | 2.110 | 2.567 | 2.898 |
| 18 | 1.330 | 1.734 | 2.101 | 2.552 | 2.878 |
| 19 | 1.328 | 1.729 | 2.093 | 2.539 | 2.861 |
| 20 | 1.325 | 1.725 | 2.086 | 2.528 | 2.845 |
| 21 | 1.323 | 1.721 | 2.080 | 2.518 | 2.831 |
| 22 | 1.321 | 1.717 | 2.074 | 2.508 | 2.819 |
| 23 | 1.319 | 1.714 | 2.069 | 2.500 | 2.807 |
| 24 | 1.318 | 1.711 | 2.064 | 2.492 | 2.797 |
| 25 | 1.316 | 1.708 | 2.060 | 2.485 | 2.787 |
| 26 | 1.315 | 1.706 | 2.056 | 2.479 | 2.779 |
| 27 | 1.314 | 1.703 | 2.052 | 2.473 | 2.771 |
| 28 | 1.313 | 1.701 | 2.048 | 2.467 | 2.763 |
| 29 | 1.311 | 1.699 | 2.045 | 2.462 | 2.756 |
| 30 | 1.310 | 1.697 | 2.042 | 2.457 | 2.750 |
| 40 | 1.303 | 1.684 | 2.021 | 2.423 | 2.704 |
| 60 | 1.296 | 1.671 | 2.000 | 2.390 | 2.660 |
| 120 | 1.289 | 1.658 | 1.980 | 2.358 | 2.617 |
| $\infty$ | 1.282 | 1.645 | 1.960 | 2.326 | 2.576 |

Source: Hoel, *Elementary Statistics,* 3rd ed. (New York: John Wiley & Sons, 1971).

## Commonly Used Standard Test Specimens

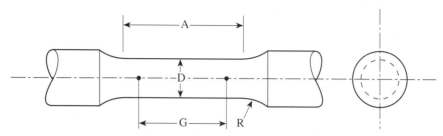

**FIGURE 3–1** Standard round tension test specimen (E 8).

where   G = gage length                        2.000 ± 0.005 in.
        D = diameter                           0.5 ± .010 in.
        R = radius                             ⅜ in., minimum
        A = length of reduced section          2¼ in., minimum

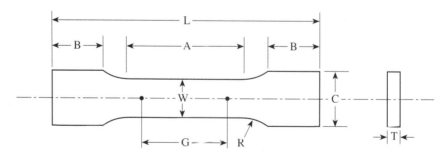

**FIGURE 3–2** Standard rectangular tension test specimen (E 8).

where   G = gage length                        2.000 ± 0.005 in.
        W = width                              0.500 ± .010 in.
        T = thickness                          thickness of material
        R = radius of the fillet               ½ in., minimum
        L = overall length                     8 in., minimum
        A = length of reduced section          2¼ in., minimum
        B = length of grip section             2 in., minimum
        C = width of grip section              ¾ in., approximate

**TABLE 3–3**  Approximate Relation of Brinell and Rockwell Hardness Numbers to Tensile Strength.

| Brinell Indentation Diameter mm | Brinell Hardness Number | | Rockwell Hardness | | Tensile Strength | |
|---|---|---|---|---|---|---|
| | Standard Ball | Tungsten Carbide ball | B Scale | C Scale | 1000 psi | MN/m$^2$ (MPa) |
| 2.45 | | 627 | | 58.7 | 347 | 2,392.22 |
| 2.5 | | 601 | | 57.3 | 328 | 2,261.23 |
| 2.55 | | 578 | | 56.0 | 313 | 2,157.82 |
| 2.60 | | 555 | | 54.7 | 298 | 2,054.41 |
| 2.65 | | 534 | | 53.5 | 288 | 1,985.47 |
| 2.70 | | 514 | | 52.1 | 274 | 1,888.96 |
| 2.75 | | 495 | | 51.0 | 264 | 1,820.02 |
| 2.80 | | 477 | | 49.6 | 252 | 1,737.29 |
| 2.85 | | 461 | | 48.5 | 242 | 1,668.35 |
| 2.90 | | 444 | | 47.1 | 230 | 1,585.62 |
| 2.95 | 429 | 429 | | 45.7 | 219 | 1,509.79 |
| 3.00 | 415 | 415 | | 44.5 | 212 | 1,461.53 |
| 3.05 | 401 | 401 | | 43.1 | 202 | 1,392.59 |
| 3.10 | 388 | 388 | | 41.8 | 193 | 1,330.54 |
| 3.15 | 375 | 375 | | 40.4 | 184 | 1,268.50 |
| 3.20 | 363 | 363 | | 39.1 | 177 | 1,220.24 |
| 3.25 | 352 | 352 | | 37.9 | 170 | 1,171.98 |
| 3.30 | 341 | 341 | | 36.6 | 163 | 1,123.72 |
| 3.35 | 331 | 331 | | 35.5 | 158 | 1,089.25 |
| 3.40 | 321 | 321 | | 34.3 | 152 | 1,047.89 |
| 3.45 | 311 | 311 | | 33.1 | 147 | 1,013.42 |
| 3.50 | 302 | 302 | | 32.1 | 143 | 985.84 |
| 3.55 | 293 | 293 | | 30.9 | 139 | 958.27 |
| 3.60 | 285 | 285 | | 29.9 | 136 | 937.58 |
| 3.65 | 277 | 277 | | 28.8 | 131 | 903.11 |
| 3.70 | 269 | 269 | | 27.6 | 128 | 882.43 |
| 3.75 | 262 | 262 | | 26.6 | 125 | 861.75 |
| 3.80 | 255 | 255 | | 25.4 | 121 | 834.17 |
| 3.85 | 248 | 248 | | 24.2 | 118 | 813.49 |
| 3.90 | 241 | 241 | 100.0 | 22.8 | 114 | 785.92 |
| 3.95 | 235 | 235 | 99.0 | 21.7 | 111 | 765.23 |
| 4.00 | 229 | 229 | 98.2 | 20.5 | 109 | 751.45 |
| 4.05 | 223 | 223 | 97.3 | | 104 | 716.98 |
| 4.10 | 217 | 217 | 96.4 | | 103 | 710.08 |
| 4.15 | 212 | 212 | 95.5 | | 100 | 689.4 |
| 4.20 | 207 | 207 | 94.6 | | 99 | 682.51 |
| 4.25 | 201 | 201 | 93.8 | | 97 | 668.72 |
| 4.30 | 197 | 197 | 92.8 | | 94 | 648.04 |
| 4.35 | 192 | 192 | 91.9 | | 92 | 634.25 |
| 4.40 | 187 | 187 | 90.7 | | 90 | 620.46 |
| 4.45 | 183 | 183 | 90.0 | | 89 | 613.57 |
| 4.50 | 179 | 179 | 89.0 | | 88 | 606.67 |
| 4.55 | 174 | 174 | 87.8 | | 86 | 592.88 |
| 4.60 | 170 | 170 | 86.8 | | 84 | 579.10 |
| 4.65 | 167 | 167 | 86.0 | | 83 | 572.2 |
| 4.70 | 163 | 163 | 85.0 | | 82 | 565.31 |
| 4.80 | 156 | 156 | 82.9 | | 80 | 551.52 |
| 4.90 | 149 | 149 | 80.8 | | 73 | 503.26 |
| 5.00 | 143 | 143 | 78.7 | | 71 | 489.47 |
| 5.10 | 137 | 137 | 76.4 | | 67 | 461.90 |
| 5.20 | 131 | 131 | 74.0 | | 65 | 448.11 |
| 5.30 | 126 | 126 | 72.0 | | 63 | 434.32 |
| 5.40 | 121 | 121 | 69.0 | | 60 | 413.64 |
| 5.50 | 116 | 116 | 67.6 | | 58 | 399.85 |
| 5.60 | 111 | 111 | 65.7 | | 56 | 386.06 |

Source: Federal Standard No. 151, Method 241.2.

## BIBLIOGRAPHY

1.  Askeland, D. R., *The Science and Engineering of Materials,* 3rd ed., PWS Publishing, 1994.

2.  Budinski, K. G., *Engineering Materials: Properties and Selection,* 4th ed., Prentice Hall, 1992.

3.  Callister, W. D., *Materials Science and Engineering: An Introduction,* 2nd ed., John Wiley, 1990.

4.  Courtney, T. H., *Mechanical Behavior of Materials,* McGraw-Hill, 1989.

5.  Dieter, G. E., *Mechanical Metallurgy,* 3rd ed., McGraw-Hill, 1986.

6.  Dowdell, R. L., et al., *General Metallography,* John Wiley, 1953.

7.  *Encyclopedia of Polymer Science and Engineering,* 2nd ed., vols. 1–17, H. F. Mark et al., eds., John Wiley, 1985–1989.

8.  *Engineering Plastics,* vol. 2, *Engineered Materials Handbook,* ASM International, 1988.

9.  Farag, M. M., *Selection of Materials and Manufacturing Processes for Engineering Design,* Prentice Hall International, 1989.

10. Flinn, R. A., and P. K. Trojan, *Engineering Materials and Their Applications,* 4th ed., Houghton Mifflin, 1990.

11. Kehl, G. L., *The Principles of Metallographic Laboratory Practice,* 3rd ed., McGraw-Hill, 1949.

12. Kingery, W. D., et al., *Introduction to Ceramics,* 2nd ed., John Wiley, 1976.

13. Krauss, G., *Principles of Heat Treatment of Steel,* ASM International, 1980.

14. Massalski, T. B., *Binary Alloy Phase Diagrams,* vols. 1 and 2, ASM International, 1986.

15. *Metals Handbook: Desk Edition,* H. E. Boyer and T. L. Gall, eds., American Society for Metals, 1985.

16. *Metals Handbook,* 9th ed., vol. 4: *Heat Treating,* American Society for Metals, 1981.

17. *Metals Handbook,* 8th ed., vol. 8: *Metallography, Structure and Phase Diagrams,* American Society for Metals, 1973.

18. *Modern Plastics Encyclopedia,* McGraw-Hill, annual.

19. Moore, G. R., and D. E. Kline, *Properties and Processing of Polymers for Engineers,* Prentice Hall, 1984.

20. Nadai, A., *Theory of Flow and Fracture of Solids,* 2nd ed., McGraw-Hill, vol. 1, 1950, vol. 2, 1963.

21.  Norton, F. H., *Elements of Ceramics,* Addison-Wesley, 1957.

22.  Phillips, V. A., *Modern Metallographic Techniques and Their Applications,* John Wiley, 1971.

23.  *Polymers Handbook,* 3rd ed., J. Brandrup and E. H. Immergut, eds., John Wiley, 1989.

24.  Reed-Hill, R. E., and R. Abbaschian, *Physical Metallurgy Principles,* 3rd ed., PWS Publishing, 1992.

25.  Schackelford, J. F., *Introduction to Materials Science for Engineers,* 3rd ed., Macmillan, 1992.

26.  Smith, W. F., *Foundation of Materials Science and Engineering,* 2nd ed., McGraw-Hill, 1993.

27.  Van der Voort, G. F., *Metallography: Principles and Practices,* McGraw-Hill, 1984.

28.  Van Vlack, L. H., *Elements of Materials Science and Engineering,* 6th ed., Addison-Wesley, 1989.

# Index